ELECTRICITY ECONOMICS

ELECTRICITY ECONOMICS

Regulation and Deregulation

GEOFFREY ROTHWELL
TOMÁS GÓMEZ

IEEE Power Engineering Society, Sponsor

IEEE Press Power Engineering Series
Mohamed E. El-Hawary, *Series Editor*

IEEE PRESS

A JOHN WILEY & SONS PUBLICATION

Permission to use material in Exercises 4.2 and 4.3 and Table 4.1 from Viscusi, W. Kip, J. Vernon, and J. Harrington. 2000. *Economics of regulation and antitrust.* 3rd ed. Cambridge: MIT Press (pages 360, 363, and 394) has been granted by the MIT Press.

Cover design adapted from Shmuel Oren (2000). Capacity payments and supply adequacy in competitive electricity markets. Presented at the VII Symposium of Specialists in Electric Operational and Expansion Planning (SEPOPE), May 21–26, Curitiba, Brazil.

Published simultaneously in Canada.

For general information on our other products and services please contact our Customer Care Department within the U.S. at 877-762-2974, outside the U.S. at 317-572-3993 or fax 317-572-4002.

Wiley also publishes its books in a variety of electronic formats. Some content that appears in print, however, may not be available in electronic format.

Library of Congress Cataloging in Publication Data is available.

Rothwell, Geoffrey.
Electricity Economics: Regulation and deregulation/
Geoffrey Rothwell and Tomás Gómez
p. cm.
Includes bibliographical references and index.
ISBN 0-471-23437-0 (cloth: alk. paper)

10 9 8 7 6 5 4 3 2 1

We dedicate this book to our families
for all their sacrifices
while we were working or drinking coffee.

CONTENTS

PREFACE

We have written this book for the thousands of professionals working in the electric utility industries around the world who need to understand the economics underlying changes in electricity regulation and emerging electricity markets. Electric utilities are undergoing profound transformations: nationally owned systems are becoming privatized, privately owned systems that were regulated are becoming deregulated, and national systems are becoming international.

The underlying theme of these changes is one of replacing monopoly with competition. Changes in electricity generation technology have prompted the realization that generation need not be a regulated monopoly to be socially efficient. Unlike transmission and distribution, which are best served by regulated monopolies under current technology, the regulation of generation and retail sales might be best done through market discipline.

Professionals in the power sector who were trained to work for electricity monopolies must now work in a new world, one in which economic efficiency is replacing technical efficiency as the cornerstone of decision making. This book is a unique attempt to provide the tools to face this new world.

We originally wrote this book as a training manual for the Federal Energy Commission of the Russian Federation to educate regional regulators. The government of Spain financed our work and the World Bank managed the project and is publishing the manual with other materials in Russian. We thank those who read and reviewed the training manual. We tried to respond to their comments. We have expanded and updated the manual to produce this book.

We assume that readers have a technical background and are not familiar with economics beyond an introductory level. So, the body of the text is presented with a minimum of mathematics. On the other hand, exercises in the chapters of the first half of the book rely on the reader's understanding of mathematics, particularly calculus. Although we suggest that the reader work through these exercises (the solutions are on our web site http://www.iit.upco.es/wiit/Electricity_Economics), the remainder of the text does not rely on the reader's understanding of the exercises. We suggest that you try the exercises. We believe that after studying this book, serious readers will understand the economic forces that are changing the international electricity industry.

Although some readers could fully understand the material through independent study, education is sometimes best accomplished in a social setting. Therefore, we have organized this book within a continuing-education context. The material can be the basis of a week-long intensive workshop in which Chapters 1–5 are covered during the first four days and one or two of the case studies (Chapters 6–9) are covered on the last day. (For this type of workshop we suggest that students read the first five chapters before the workshop.) The material could be presented in a 10-week quarter with one week on each chapter. Or the material could be presented in a 15-week semester with (1) two weeks on each of the early chapters (2–5), (2) a week on each case study, (3) the assignment of a paper applying our case study outline to a particular electricity system, and (4) presentation of these papers in the last weeks of the semester. Whether the material is studied independently or in a more formal context, we wish you luck and encourage you to email us with your comments.

ACKNOWLEDGMENTS

We extend our gratitude to Gary Stuggins of the World Bank for his support in producing the first draft of this book as a training manual for the Federal Energy Commission of the Russian Federation and for his assistance in seeking permission to publish an expanded and updated version. We thank Prof. J. Ignacio Pérez-Arriaga of the Universidad Pontificia Comillas, Prof. Shmuel Oren of the University of California, Berkeley, and anonymous reviewers for their insightful comments on the contents and organization of the book. Ramón Sanz from Mercados Energéticos in Argentina kindly reviewed Chapter 9 and Fernando Perán from Comillas helped collect data for the Argentine case study. We would like to acknowledge the help received from Prof. Hugh Rudnick, Pontificia Universidad Católica de Chile, who suggested that we publish this with IEEE press. We thank colleagues at Instituto de Investigación Tecnológica, especially Carlos Vázquez and Michel Rivier; at University of California, Berkeley, including Richard Gilbert and Severin Borenstein; and at Stanford University, including Tim Bresnahan, Roger Noll, and Frank Wolak, for interesting discussions and work on electricity deregulation. Finally, we thank all of those at the IEEE press who made this publication possible, particularly Robert Bedford and Anthony VenGraitis.

Geoffrey Rothwell thanks the authors for their patience with his editing, the authors of other textbooks and educational materials for their guidance and comments, including Darwin Hall, Joseph Harrington, Dan Rubinfeld, Bob Pindyck, Steven Stoft, John Vernon, and Kip Viscusi. Also, he thanks the participants (particularly, Lucille Langlois) of the International Atomic Energy Agency's "Eastern European Training Workshop on Analysis of Economic and Financial Aspects of Nuclear Power Programmes," Zagreb, Croatia (May 11–14, 1998), who provided an implicit audience.

Tomás Gómez thanks Chris Marnay, Joe Eto, and Ryan Wiser for their friendship and support during his sabbatical stay at the Ernest Orlando Lawrence Berkeley National Laboratory (LBNL) in California. Also, he thanks the Spanish Ministry of Education and Culture for its financial support and the Universidad Pontificia Comillas that made possible the sabbatical leave to write this book.

GEOFFREY ROTHWELL
TOMÁS GÓMEZ

Stanford, California
Madrid, Spain
January 2003

NOMENCLATURE

ACRONYMS AND ABBREVIATIONS

AB	Assembly Bill (California)
AGC	Automatic Generation Control
bbl	Barrels
B	Billions
BOE	Boletín Oficial del Estado (Spain)
Btu	British thermal unit
CAPM	Capital Asset Pricing Model
CC	Capacity Charge
CCGT	Combined-Cycle Gas Turbine
CD	Cost Drivers
CfD	Contract for Differences
CFR	Code of Federal Regulations (U.S.)
CO_2	Carbon Dioxide
COS	Cost-of-Service (regulation)
CTC	Competition Transition Charge (e.g., California)
CTTC	Costs of Transition to Competition (Spain)
Co.	Company
DASR	Direct Access Service Request
DC	Direct Current
DEA	Data Envelopment Analysis
DISCO	Distribution Company
DKK	Denmark Krone (Danish currency)
DSM	Demand Side Management
ENS	Energy Not Served or Energy Not Supplied
EPAct	Energy Policy Act of 1992 (U.S.)
ESP	Energy Service Provider
EURO	European Union currency
EWG	Exempt Wholesale Generators
FDC	Fully Distributed Costs
FIM	Finish Mark (currency)

FTR	Firm Transmission Rights
GDP	Gross Domestic Product
GW	Gigawatt
GWh	Gigawatt-hour
HV	High Voltage
IOU	Investor-Owned Utility
IPP	Independent Power Producers
ISO	Independent System Operator
IVA	Independent Verification Agent
k	Kilo (thousand)
kV	Kilovolts (thousands of volts)
kW	Kilowatts (thousands of watts)
kWh	Kilowatt-hours (thousands of watts per hour)
LOLP	Loss of Load Probability
LOSEN	Ley Orgánica del Sistema Eléctrico Nacional (Spain)
LV	Low Voltage
M	Millions
MBtu	Millions of Btus
MCP	Market Clearing Price
MDMA	Meter Demand Management Agent
MLE	Marco Legal Estable (Spain)
MO	Market Operator
MSP	Meter Service Providers
MV	Medium Voltage
MW	Megawatt
MWh	Megawatt-hour
NOPR	Notice of Proposed Rulemaking (U.S.)
NETA	New Electricity Trading Arrangements (UK)
NIEPI	Average System Interruption Frequency Index (Spain)
NOK	Norwegian Krone (currency)
NOPR	Notice of Proposed Rulemaking (U.S.)
NUG	Non-Utility Generator
O&M	Operation and Maintenance (costs)
OASIS	Open Access Same-time Information System
OTC	Over-the-Counter (market)
PBR	Performance-Based Ratemaking
poolco	A pool structure in which all generators must sell to the pool
PUCHA	Public Utility Holding Company Act (U.S.)
PURPA	Public Utilities Regulatory Policy Act (U.S.)
Pta	Peseta (Spanish currency)
QF	Qualifying Facilities
RD&D	Research, Development, and Demonstration
RF	Retribución Fija (stranded cost recovery payment in Spain)
RMR	Reliability Must Run
ROR	Rate-of-Return (regulation)

SA	Société Anonyme (a publicly held company, similar to incorporated)
SALEX	Funds to finance Argentina transmission system
SEK	Swedish Krona (currency)
SO	System Operator
TIEPI	Average System Interruption Duration Index (Spain)
TRANSCO	Transmission Company
TWh	Terawatt-hour
Tcf	Trillions of cubic feet (of natural gas)
UDC	Utility Distribution Company
VOLL	Value of Loss of Load

ORGANIZATIONS AND ENTITIES (Also see Table 3.2)

ADEERA	Argentine Distributors Association
AGEERA	Argentine Generators Association
ATEERA	Argentine Transmitters Association
Alicurá	A hydroelectric plant in Argentina
AyEE	A utility in Argentina
CAM	Comité de Agentes del Mercado (Market Agents Committee, Spain)
CAMMESA	Compañía Administradora del Mercado Eléctrico Mayorista SA (Argentina)
CEC	California Energy Commission
CENELEC	European Committee for Electrotechnical Standardization
CNSE	Comisión Nacional del Sistema Eléctrico (Spain)
CPUC	California Public Utilities Commission
Central Costanera	A power plant in Argentina
Central Puerto	A power plant in Argentina
Chevron	A U.S. corporation
Chocón	A hydroelectric power plant in Argentina
Citigroup	A U.S. corporation
Detroit Edison	A U.S. electric utility
DISGRUP	Grupo de Trabajo de Distribucíon y Comercializacíon (Spain)
DISTRELEC	A distribution company in Argentina
DOJ	Department of Justice (U.S.)
Dow Jones	A U.S. corporation
EC	European Community
ECON	Center for Economic Analysis (Norway)
EDELAP	A distribution company in Argentina
EDENOR	A distribution company in Argentina
EDESUR	A distribution company in Argentina
EDF	Electricité de France

EIA	Energy Information Administration (U.S.)
EL-EX	Finnish Electricity Exchange Market
ELBAS	Joint Swedish/Finnish Adjustment Market
ENEL	Ente Nacional de l'Energia (Italy)
ENRE	Ente Nacional Regulador de la Electricidad (Argentina)
ERZ	A generating company in Spain
ESEBA	A power plant in Argentina
EU	European Union
Elkraft	A transmission system operator in Eastern Denmark
Elspot	Norwegian Electricity Spot Market
Eltermin	Norwegian Futures and Forwards Market
Eltra	A transmission system operator in Western Denmark
Endesa	A generation holding company in Spain
Enher	A generating company in Spain
Euro Cable	A transmission line between Norway and Germany
FERC	Federal Energy Regulatory Commission (U.S.)
FPC	Federal Power Commission (precursor to FERC)
FTC	Federal Trade Commission (U.S.)
Fecsa	A generating company in Spain
Fingrid	National grid operator of Finland
GUMA	Large Users Association (Argentina)
GUME	Minor Large Users of Electricity (Argentina)
GUPA	Particular Large Users of Electricity (Argentina)
Gesa	A generating company in Endesa Holding (Spain)
Hecsa	A generating company in Endesa Holding (Spain)
Hidronor	A former state-owned utility (Argentina)
IEEE	Institute of Electrical and Electronic Engineers
Iberdrola	A generating company in Spain
LADWP	Los Angeles Department of Water and Power, a public utility
LBNL	Ernest Orlando Lawrence Berkeley National Laboratory
MEM	Mercado Electrico Mayorista (Argentina)
MEOSP	Ministerio de Economía y Obras y Servicios Públicos (Argentina)
Moody's	Moody's Investors Service, a U.S. corporation
NARUC	National Association of Regulatory Utility Commissioners (U.S.)
NERA	National Economic Research Associates (U.S.)
NGC	National Grid Company (UK)
NORDEL	Nordic Organization for Electric Cooperation
NVE	Norwegian Water Resources and Energy Directorate
NYMEX	New York Mercantile Exchange
NYSE	New York Stock Exchange
NorNed	A cable connecting Norway and the Netherlands
Nord Pool	Norwegian electricity market operator
OED	Norwegian Ministry of Petroleum and Energy

OMEL	Compañía Operadora del Mercado Español de Electricidad (Spain)
OPEC	Organization of Petroleum Exporting Countries
PAFTT	Technical Transmission Function Providers (Argentina)
PG&E	Pacific Gas & Electric, U.S. corporation
PJM	Pennsylvania–New Jersey–Maryland, an electricity pool (U.S.)
PX	Power Exchange (California)
P. del Aguila	A hydroelectric plant in Argentina
Palo Verde	An interchange on the Arizona-California border
REE	Red Eléctrica de España SA (Spain)
SADI	Interconnected Transmission System of Argentina
SCE	Southern California Edison, U.S. corporation
SDG&E	San Diego Gas & Electric, a U.S. corporation
SEC	Securities Exchange Commission (U.S.)
SEGBA	Servicios Eléctricos del Gran Buenos Aires, SA
SK	Svenska Kraftnat, Swedish national power grid
SMUD	Sacramento Municipal Utilities District, a public utility in California
Salto Grande	A power plant in Argentina
Samkjøringen	An early Norwegian power pool
Sevillana	A generating company in Spain
Smestad	An Oslo area bilateral forward contracts reference price
SSB	Statistics Norway
Statkraft	A large state-owned generator in Norway
Statnett SF	Norwegian National Grid Company
Storebaelt	A Danish interconnection
SwePol	An interconnection between Sweden and Poland
TRANSENER SA (Argentina)	Compañía de Transporte de Energía Eléctrica en Alta Tensíon
UCTE	Union for the Coordination of the Transmission of Electricity (Europe)
Unelco	A generating company in Endesa Holding (Spain)
Union Fenosa	A generating company in Spain
Viesgo	A generating company in Endesa Holding (Spain)
Viking Cable	A cable connecting Norway and Germany
WEPEX	Western Power Exchange (U.S.)
Yacilec	A transmission company in Argentina
Yacyreta	A hydroelectric power plant in Argentina

MATHEMATICAL SYMBOLS (Also see Table 9.10)

(Note $ is used to represent values that take currency as a unit.)

a	A parameter
α	Capital Asset Pricing Model parameter

A	Annuity ($)
AC	Average Cost ($)
AFC	Average Fixed Cost ($)
AIC	Average Incremental Cost ($)
AR	Average Revenue ($)
ARA	Absolute Risk Aversion
AVC	Average Variable Cost ($)
b	A parameter
β	Capital Asset Pricing Model parameter
c	A parameter
C	Cash flow ($)
CC_H	Capacity Charge in Highly congested zone ($, Norway)
CC_L	Capacity Charge in Lowly congested zone ($, Norway)
CEQ	Certainty Equivalent ($)
CF	Capacity Factor (%)
CGA	Customer Growth Adjustment factor ($)
$CORR$	Correlation (statistics)
COV	Covariance (statistics)
CPI	Consumer Price Index (%)
CRF	Capital Recovery Factor (%)
CS	Consumer Surplus ($)
Ce	Consumption of energy (kWh)
d	A parameter
d	Derivative (calculus)
Δ	Rate of change
D	Demand function or curve
DR	Distribution Company Revenue ($)
Dc	Distribution costs ($)
e	Exponential function (mathematics)
E_d	Elasticity of demand
E_s	Elasticity of supply
E	Expectation operator (statistics)
EFK	Efficiency improvement factor (%, Norway)
Eff	Efficiency factor (%, Spain)
f	Price of fuel (e.g., $/MBtu)
F	Fuel (e.g., in MBtu)
FC	Fixed Cost ($)
FV	Future Value ($)
Fc	Fixed Charge ($)
g	A parameter
GC	Commercial costs per user ($)
GN	Generator
H	Trigger value to invest ($)
i	Inflation (%)
I	Investment ($)
IIN	New transmission investments ($, Spain)

IPC	Retail price index (%, Spain)
IRR	Internal Rate of Return (%)
IT	Revenue cap ($)
j	Index indicator
k	Index indicator
K	Capital ($)
KPI	Consumer price index (%, Norway)
KR	User demand responsibility factor (Argentina)
ln	Natural logarithm
L	Labor (e.g., in hours)
LE	Growth factor (Norway)
min	Minimization function (mathematics)
MC	Marginal Cost ($)
MR	Marginal Revenue ($)
MRTS	Marginal Rate of Technical Substitution
n	Total number
NPV	Net Present Value ($)
OLS	Ordinary Least Squares (statistics)
P	Price ($)
PR	Profit ($)
PS	Producer Surplus ($)
PV	Present Value ($)
Q	Quantity (e.g., MWh)
r	Real rate of interest (%)
r_{debt}	Real rate of interest on debt (%)
r_{equity}	Real rate of return on equity (%)
r_f	Real risk-free rate of interest (%)
r_m	Rate of return on a market portfolio (%)
R	Nominal rate of interest (%)
R_f	Nominal risk-free rate of interest (%)
RB	Rate Base ($)
RP	Risk Premium ($)
RPC	Revenue Per Customer ($)
RPI	Retail Price Index (%)
RR	Required Revenues ($)
RRA	Relative Risk Aversion
s	Allowed rate of return (%)
SD	Standard Deviation (statistics)
t	Time index (e.g., year)
T	Final period in a sequence
TC	Total Cost ($)
TR	Total Revenues ($)
U	Utility function or level of utility
UOP	Annualized operating cost per unit of power capacity ($, Argentina)
V_W	Value of Waiting ($)

V	Value of a project ($)
VAR	Variance (statistics)
Vc	Variable Charge ($)
VC	Variable Cost ($)
w	Wage rate (e.g., $/hour)
W	Wealth ($)
$WACC$	Weighted Average Cost of Capital (%)
x	A variable
X	Productivity or efficiency offset factor in incentive regulation (%)
y	A variable
z	A variable
Z	Exogenous influences on utility under PBR ($)

CHAPTER 1

ELECTRICITY REGULATION AND DEREGULATION

1.1. THE ELECTRICITY INDUSTRY: RESTRUCTURING AND DEREGULATION

During the 1990s, a deep transformation in the electricity industry took place in many countries. This sector is moving from a monopoly structure to a more competitive one, as are the transportation and telecommunications sectors. For example, in Latin America, Chile was a pioneer in the early 1980s with the development of a competitive system for electricity generation based on marginal prices. In 1992, Argentina privatized an inefficient government-owned electricity sector, splitting it into generation, transmission, and distribution companies, and introduced a competitive generation market (see Chapter 9). These experiences were repeated in other countries in the region, such as Bolivia, Peru, Colombia, Guatemala, El Salvador, Panama, and, to a limited extent, Brazil and Mexico.

In Europe, Scotland and Northern Ireland followed the experience of England and Wales (Littlechild and Beesley, 1989). The Scandinavian countries, following Norway, have gradually created a Nordic wholesale electricity market (see Chapter 7). In the European Union in 1996, the European Parliament and Council issued the Internal Electricity Market Directive 96/92/EC that set goals for a gradual opening of national electricity markets and rules for transmission access in the 15 member states (European Parliament and Council, 1996; Schwarz, Staschus, Knop, and Zettler, 2000). Spain in 1998 and Netherlands in 1999 created fully competitive generation markets (see Chapter 8). The rest of the members are adapting to the new regulations. For an international comparison of transmission grid access, see Grønli, Gómez, and Marnay (1999). For other international comparisons, see Gilbert and Kahn (1996).

In New Zealand, Australia, and some provinces of Canada (Alberta and Ontario), deregulation of the electricity industry is being introduced as a way of increasing efficiency and reducing prices. This is also true in some states of the United States (US); restructuring legislation has already been enacted in half the states, with California and Pennsylvania–New Jersey–Maryland (PJM) in the lead. See Stoft (2002). However, the California electricity crisis of 2000 and 2001 has slowed the move toward electricity deregulation in the United States (see Chapter 6).

Under restructuring and deregulation, vertically integrated utilities, in which producers generate, transmit, and distribute electricity, have been legally or functionally unbundled. Competition has been introduced in the wholesale generation and retailing of electricity. Wholesale electricity markets are organized with several generation companies that compete to sell their electricity in a centralized pool and/or through bilateral contracts with buyers. Retail competition, in which customers can choose among different sellers or buy directly from the wholesale market, has also been implemented. This was done instantaneously for all customers (as in Norway), or progressively, under a multiyear program, according to different customer sizes (as in England and Wales, Australia, Argentina, etc.).

Transmission and distribution are still considered natural monopolies (see Chapter 2 for the economics of monopolies and natural monopolies) that require regulation (see Chapter 4 on the regulation of natural monopolies). To achieve effective competition, regulation is still needed to ensure open, nondiscriminatory access to the transmission grid for all market participants.

1.2. FROM MONOPOLIES TO MARKETS

Restructuring and deregulation involve a transformation in the structure and organization of electricity companies. Traditionally, a single utility, vertically integrated, was the only electricity provider in its service territory and had the obligation to supply electricity to all customers in its territory. This provider could be

- Owned by a national, regional, or local government
- Owned by a cooperative of consumers
- Owned privately

Because of the monopoly (single seller) status of the provider, the regulator periodically sets the tariff to earn a fair rate of return on investments and to recover operational expenses; see Chapter 3 on determining the rate of return and Chapter 4 on rate-of-return regulation. Under this regulated framework, firms maximize profit subject to many regulatory constraints. But because utilities have been allowed to pass costs on to customers through regulated tariffs, there has been little incentive to reduce costs or to make investments with due consideration of risk.

Under perfect competition, in theory, the interaction of many buyers and sellers yields a market price that is equal to the cost of producing the last unit sold. This is the *economically efficient* solution. The role of deregulation is to structure a competitive market with enough generators to eliminate ***market power*** (i.e., the ability of a firm or a group of firms to set prices "a small but significant and non-transitory amount" above production cost; see DOJ/FTC, 1992). (See the Glossary for the definitions of words and phrases in ***bold italic*** type.)

With deregulation, electric utilities must split regulated from deregulated activities and compete with new firms originating from other energy businesses or retail services (see Chapter 5). The economic decision-making mechanism, under competition, responds to a decentralized process whereby each participant maximizes profit equal to the difference between total revenue and total cost. However, under competition, the recovery of investment in new plant is not guaranteed. So, risk management becomes a crucial part of the electricity business.

1.3. WHY RESTRUCTURING AND DEREGULATION NOW?

There are many forces driving electricity restructuring around the world. These forces are

1. New generation technologies, such as combined-cycle gas turbines (CCGT), have reduced the optimal size of an electricity generator.
2. The competitive global economy requires input cost reduction; electricity is a primary input for many industries.
3. The State, as owner and manager of traditional infrastructure industries, cannot respond as quickly as private owners to economic and technological change, prompting privatization.
4. Information technologies and communication systems make possible the exchange of huge volumes of information needed to manage electricity markets.

CCGT manufacturers have been racing to achieve (1) technical efficiencies close to 60%, (2) short power plant construction periods (less than 2 years), and (3) low investment costs (around U.S.$500/kW). These technical developments (along with low natural gas prices and new natural gas transportation networks) have made this technology the dominant choice for new investment in competitive generation markets.

Even before the opening of generation to competition, CCGT technology was being built by independent power producers selling electricity to traditional utilities under different types of regulated agreements. The efficient size of these power units is currently between 150 and 300 MW. This is much smaller than efficient scales for traditional fossil or nuclear power stations.

Global competition promoted by international firms is emphasizing international price comparisons and, consequently, inducing nations to reduce electricity costs to be globally competitive. Restructuring and deregulation processes are carried out by governments through the introduction of electricity markets to increase efficiency and reduce prices. Markets also promote participation of external agents and neighboring countries with lower production costs as a way to achieve lower prices.

After World War II, in many countries, for strategic reasons, the electricity industry was gathered in a single, nationalized company. This situation was common in Europe and Latin America. But public ownership has been in crisis during the last decade for various reasons. For instance, in Latin American countries that had high rates of electricity demand growth, the State, with a significant external debt, was unable to carry out the needed generation investments. This situation, plus the recommendations of international financial institutions, such as the World Bank and the Inter-American Development Bank, led governments to initiate privatization and restructuring.

Also, the internationalization of fuel markets called into question national subsidies to specific primary energy sources. For instance, in several countries in Europe, the State has been subsidizing the coal industry. Low international coal prices (and the usual environmental problems associated with burning low-quality domestic coal) prompted governments to progressively abandon this type of intervention. Similarly, the nuclear power industry was developed with a high level of State support. However, political opposition has undercut this support, postponing or stopping new investment in nuclear plants.

Finally, information technologies and communication systems are making possible day-ahead and on-line electricity markets with multiple agents and multiple types of transactions. Further, metering, billing, quality control, and load management options based on new information technologies and communication systems are being offered under restructuring and deregulation. Also, retail competition and customer choice based on these technologies encourages entry of new electricity service providers with new commercial relationships, offering attractive prices, high quality, and other integrated services.

1.4. REGULATION IS STILL REQUIRED

Although regulators' objectives differ across countries and sectors, their primary objective is to protect the short-run and long-run interests of consumers by promoting economic efficiency. The most direct way to achieve efficiency is to encourage or mimic competition. However, economic regulation must be used where competition is not feasible, for example, in sectors that have natural monopoly characteristics or in situations where externalities have not been internalized.

Traditionally, the electricity industry has been dominated by monopolies. Under restructuring, only high-voltage transmission, distribution, and system operation exhibit natural monopoly characteristics. Achieving economic efficiency in natural monopoly industries requires regulation. In these industries, the largest firms can charge the lowest prices, driving rivals from the market. Once there is no competition, the surviving firm can charge monopoly prices, reducing quantity and social welfare (see Chapter 2). There are several solutions to this problem, including

1. Government ownership of the industry, with a mandate to provide adequate output at reasonable prices
2. Private ownership with government regulation to ensure adequate output and a reasonable return on private investment

The economic theory of regulation (see overview in Joskow and Noll, 1981) attempts to predict which institutional arrangement is preferable as a function of the comparative social costs and benefits of

- Private monopoly without regulation
- Government monopoly
- Private monopoly with regulation

Each solution involves costs, including (1) the social cost of the monopolist using its market power, (2) the cost of maintaining a regulatory agency, and (3) the costs imposed on the monopolist by the regulator. Besides the administrative costs associated with regulation, another potential cost arises from misguided regulatory interventions that can create social welfare losses. Therefore, the regulator must carefully consider the costs and benefits of each regulatory requirement on the regulatory agency and the regulated utility (see Chapter 4).

The role of regulation is to encourage enough investment to meet customer demand and to compensate investors with a reasonable rate of return. There are several ways of accomplishing regulatory goals in the electric power industry. Two basic regulatory forms are (1) *Rate-of-Return* (ROR), also known as *Cost-of-Service* (COS), regulation, which requires the regulator to actively monitor the electric utility; and (2) *Performance-Based Ratemaking* (PBR), which requires much less regulator intervention. Under ROR or COS regulation the regulator determines

1. Appropriate expenses
2. The value of invested capital
3. The allowed rate of return on invested capital

This process requires a costly exchange of information between the regulator and the electric utility. PBR involves mechanisms that attempt to reduce the cost of regulation by allowing utilities to keep profits resulting from efficient operation.

As an electricity industry is restructured, the role of the regulator becomes one of setting market guidelines to yield competitive conditions in which prices and quantities are similar to what they might be under perfect competition (see Chapter 5).

Establishing competitive electricity markets requires a reduction in the market power that could be exerted by the formerly integrated utilities. In some cases, the regulator has obliged these utilities to divest their generation assets. Economic efficiency gains from deregulation can disappear if there is no real competition at the wholesale level.

On the other hand, usually, retail competition is initially dominated by the utilities that formerly distributed electricity to customers. They can also create their own retail or service provider companies as deregulated firms. The role of the regulator in this area is crucial to ensure fair competition. Regulated distribution companies, as former vertically integrated utilities, will provide preferential treatment to their own spin-off retailers rather than to new entrants. The regulator should establish clear rules to avoid this discriminatory behavior, while actively promoting the entrance of new participants.

Where regulation is maintained or introduced after privatization, regulators should adopt open, transparent, and objective decision-making procedures (i.e., observable data sources, replicable methods, open debate, and reasoned decisions). This is because regulatory decisions are always a part of an ongoing regulatory regime. Electricity companies will continue to be regulated where capital-intensive investments can lead to monopoly conditions. In the current environment, these conditions clearly apply to investments in transmission and distribution.

In a regulatory regime that sets revenue for an industry characterized by assets with long lives, the credibility of regulatory commitments is extremely important. Before investors will commit funds to such investments, they must be convinced that the regulator will allow future revenues that provide reasonable assurance of cost recovery. For example, preventing the recovery of *stranded costs* or *assets* (see Chapter 5) associated with past investments would allow the regulator to make an immediate price reduction, but also reduces the necessary credibility that future investments might be recovered. Therefore, the regulator must consider both consumers' short-run interests in low-price, high-quality service and their long-term interests in continued maintenance and investment in the electric power sector.

1.5. WHAT LESSONS CAN BE LEARNED FROM INTERNATIONAL EXPERIENCES?

The economics of natural monopolies, markets, and regulation are not enough to understand the complexities of real regulatory reforms. There are many issues of practical implementation that should be analyzed through case studies to obtain a clearer understanding of electricity restructuring. For that purpose, we have selected four restructuring experiences to describe in detail in this book. These experiences correspond to the cases of California, Norway, Spain, and Argentina. This complexity is portrayed in Table 1.1, which compares the institutional, regulatory, organizational, and technical issues of the four case studies.

We begin with California because of the problems during its transition from regulation to deregulation. In the late 1990s, California tried to ensure compatibility between *bilateral trading* and a *centralized pool.* In addition, California addressed the issue of the stranded costs of the former investor-owned regulated utilities. To recover these stranded costs, electricity tariffs were frozen at a regulated tariff 10% below 1996 levels and a *competition transition charge* was added to them. Consequently, when the stranded costs were recovered and regulated tariffs disappeared, customers faced the high prices of the wholesale market. See Chapter 6 on this and other issues associated with the California "electricity crisis."

Norway's original restructuring design was one of a wholesale, competitive market based on bilateral trading. However, it was extended to incorporate international trading with other Scandinavian countries and retail access with the opening of the market to small customers. These are characteristics that justify the inclusion of this case. Further, markets for peak power, including demand-side bidding, are under development in Scandinavia (see Chapter 7).

Electricity restructuring in Spain is similar to California's wholesale market design and stranded cost recovery. However, the starting point before deregulation in

Table 1.1. Electricity Restructuring Reforms in California, Norway, Spain, and Argentina

	California	Norway	Spain	Argentina
Conditions driving restructuring	• State's electricity higher than national average • Federal regulation: EPAct (1992), FERC orders (1996) • Inefficient centralized regulation	• Promote efficiency in investments and reduction of regional price differences • Avoid cross-subsidization among customer groups • Create cost reduction incentives	• Electricity prices higher than prices in neighboring countries • European Directive (1996) • Eliminate subsidies to the coal industry	• Power shortages due to lack of investment and generation unavailability • Highly inefficient public owned sector • Need for new investment
Restructuring law	AB 1890 (1996)	Energy Law (1990)	Electricity Law (1997)	Electricity Law (1992)
Structural changes	• Functional separation of Generation, Transmission, and Distribution (G, T, D) • Recovery of stranded costs • Generation divestiture	• Accounting separation of regulated and competitive functions • The transmission grid was separated as a new company • No privatization of publicly owned sector	• Legal separation of regulated and competitive functions • Recovery of Endesa • Privatization of stranded costs (main publicly owned generator)	• Vertical (G, T, D) and horizontal disintegration • Privatization of federal and provincial companies

(continues)

Table 1.1. Electricity Restructuring Reforms in California, Norway, Spain, and Argentina *(continued)*

	California	Norway	Spain	Argentina
Regulatory and market institutions	• FERC and CPUC (federal and state regulators) • ISO (system operator)	• NVE (regulator) • Statnett (grid owner and system operator) • Nord Pool (market operator)	• MIE and CNSE (regulators) • REE (system operator and transmission grid owner) • OMEL (market operator)	• Secretary of Energy and ENRE (regulators) • CAMMESA (system and market operator)
Wholesale market	• Centralized and physical bilateral trades • Several transmission owners	• Centralized and physical bilateral trades • Trading in the Nordic Pool	• Centralized and physical bilateral trades	• Mandatory pool with financial bilateral contracts
Retail competition and customer choice	• All customers (1998) • Metering and billing competition	• All customers (1991)	• Gradual implementation in a 5-year period • All customers in 2003	• Large users (1992), and small customers in the future

Spain was different from California's. For several years, a single independent company in Spain was operating as transmission owner and operator. This facilitated the introduction of a wholesale market. In addition, the previous regulatory framework in Spain set a national benchmark for efficiency whereby utilities were regulated in competition by comparison (see Chapter 8).

Finally, Argentina is an example of a privatization resulting in a competitive wholesale electricity market with new generation investment during the last decade. Regulatory reforms in Argentina and Chile have influenced all other reforms in Latin America (see Chapter 9).

Other experiences are also relevant, such as those of England and Wales, Australia, and the PJM (Pennsylvania–New Jersey–Maryland) in the United States. We do not include case studies of these experiences. They are left as exercises for our readers and we encourage them to submit cases to our web site www.iit.upco.es/witt/Electricity_Economics. Also, because international experiences become outdated with the fluid state of deregulation and restructuring, at the end of the book we list World Wide Web sites where it is possible to find more up-to-date documentation.

Many organizational, institutional, and regulatory issues must be solved with deregulation. (For discussions of specific issues, see Ilic, Galiana, and Fink, 1998.) Although the ultimate objective is to achieve a technically reliable and financially viable competitive electricity supply industry, each government has adopted different approaches to restructuring. In the remainder of this chapter, we review the motivations that led to restructuring and the solutions adopted to address transitional issues. This discussion serves as an introduction to the case studies in Chapters 6–9.

1.5.1. Starting Points and Motivations for Deregulation

A combination of factors promotes the political will to deregulate. Nationally owned systems have been segregated into different companies and then privatized under a new regulatory competitive framework. This is the case for the experiences in Argentina, Chile, and England and Wales, where the ideology of the government was clearly oriented toward a general liberalization program in the country. In Argentina, in addition, the situation of a chronic lack of investment, high growth in demand, and frequent power outages, encouraged the adoption of dramatic changes.

Electricity prices higher than those in neighboring countries or regions have also pushed deregulation. In high-price areas, customers and governments influenced by a general wave of deregulation have advocated restructuring. For example, Spain was encouraged by European Directive 96/92/EC that called for the introduction of competition. In both Spain and California, the electricity industry was primarily private before restructuring. Therefore, privatization was not an issue.

However, another issue arises when private, regulated utilities have expected *required revenues* that are greater than what they would be in a competitive market. This difference is known as ***stranded cost.*** A recovery procedure for stranded cost can be designed by the regulator and used during a transition period. Also, where investor-owned utilities are required to divest their generation assets to mitigate possible market power problems, the difference between the book value of these generating assets and the price received for them in the market is known as ***stranded assets***. We will refer to both stranded costs and stranded assets as "stranded cost."

Another objective pursued by deregulation is to avoid cross-subsidies among different customer classes by designing more transparent tariffs. Electricity is bought in the market at posted prices, whereas regulated costs (e.g., for transmission services) are charged under a separate system through access tariffs. Additionally, under deregulation, subsidies to domestic primary fuels, such as coal, and to nuclear power, progressively disappear, as in Spain and in England and Wales.

1.5.2. Structural Changes and System Operation

The introduction of electricity competition requires the separation of competitive from still-regulated functions. In most restructuring experiences, the transmission grid has been separated, in ownership and in operation, from generation companies by creating a regulated transmission owner and operator. This is the case in England and Wales, Argentina (with separation between system operator and transmission owner), Norway, and Spain.

However, the situation is more complicated in California and other US states in which utilities have retained ownership of some generation assets and parts of their transmission grid. Here, new entities have been created to control the operation of the interconnected transmission grid. This is an attempt to prevent a utility from manipulating its grid to the disadvantage of competing generators.

Another key regulatory issue concerning system operation is how to maintain reliable operation under the unbundled structure. Regulated, vertically integrated utilities cooperated voluntarily to operate a reliable system by coordinating their resources with neighboring utilities, knowing that regulated tariffs would cover bundled costs. Under deregulation, the *system operator* is responsible for system reliability. It buys different *ancillary services* from generators and users to maintain a reliable system. However, legal responsibilities of system operators (particularly those that do not own transmission assets) must be clearly defined by new regulations.

On the other hand, transmission grids were not designed to transmit power flows from electricity markets. To do so requires updating transmission planning procedures and defining transmission investment responsibilities between system operators and transmission owners. This is especially true in those cases in which these functions have been separated, as in Argentina and throughout the US. Further, systems with transmission congestion problems use locational prices as a mechanism of sending market participants the right economic signal for using congested paths. In that sense, market participants can promote grid investments according to the economic value they perceive. Chile, Argentina, and PJM have *nodal prices*, whereas Norway and California have *zonal prices*. See Chapter 5 for more on nodal and zonal prices.

Therefore, under deregulation, transmission and distribution, also known as "wires businesses," continue to be regulated. *Performance-Based Ratemaking*

(PBR) regulation is being introduced through price or revenue caps that limit company revenues during a regulatory period of several years (see Chapter 4). England and Wales were the first to experience price caps as a formula to remunerate regulated activities performed by distribution companies. Argentina, California, Norway, and Spain also use PBR formulas for the same purpose. In addition, associated with the concern that cost reductions can lead to quality degradation, mechanisms to control service quality are also being used.

1.5.3. Design of Wholesale Markets and Market Institutions

A major objective of electricity deregulation is to achieve a workably competitive wholesale market. See Stoft (2002, Part 3). At first, wholesale markets were designed for economic dispatch of generating units in a centralized pool, managed by the system operator. Participation in the pool was mandatory for all generators. This was the case in Argentina, Chile, and England and Wales. (For an international comparison of power pool operations, see Barker, Tenenbaum, and Wolf, 1997.) Generators declared costs, or submitted bids, to the system operator who (using economic dispatch algorithms) obtained the generation schedule and hourly marginal prices (in England and Wales for each half hour). There was no demand-side bidding. Also, in Chile and in Argentina long-term marginal prices (3 to 6 months), instead of hourly prices, were passed through to regulated final customers. Unregulated customers could buy electricity with financial contracts.

In Norway, however, the wholesale market design was based on bilateral bidding with both generation and demand bids. A market for *futures contracts* (up to 3 years in advance) was also instituted. Market operations were coordinated by a separate entity distinct from the system operator, specifically created for this purpose—the market operator. Later, as in California and Spain, market operations were separated from system operations. Energy transactions can be made in a centralized pool or directly, outside the pool, through bilateral contracts.

Wholesale electricity markets have high price volatility due to daily and seasonal variations in supply and demand. This raises two important issues under deregulation: demand responsiveness to price variations and new investment in generation resources.

Under regulation, electricity demand was considered inelastic and new capacity was built to cover projected demand to minimize investment plus operating costs. Under deregulation, it is assumed that competitive prices will encourage new generation. In some cases (e.g., in Argentina, Chile, England and Wales, and Spain) besides energy revenues obtained from selling electricity, generators are paid a supplemental capacity payment to encourage generation investment. In other cases (e.g., Australia, California, New Zealand, and Norway), this supplemental payment is not used.

This is a controversial issue. In Argentina, generation investment has been successful, even when wholesale prices are depressed, because of capacity payments. On the other hand, California has experienced high price volatility and high average wholesale prices due to high fuel prices and delays in generation investment. In

Scandinavia, markets for peak power (capacity markets) are under development. Elsewhere, there are proposals to address the issue of long-term electricity supply. For example, by using market mechanisms, consumers and generators can arrange long-term contracts so consumers can cover their expected needs and generators can stabilize incomes to recover fixed investment costs.

1.5.4. Retail Competition and Customer Choice

The aim of deregulation is to provide market-based electricity prices to customers with reliable service at efficient prices. Wholesale competition is enhanced, on the supply side, by participation of several generation firms, and, on the demand side, by allowing customers to buy directly or indirectly from generators through customer choice and retail competition.

The introduction of customer choice differs from country to country. In Norway, all customers were ***qualified*** to choose their supplier when the competitive wholesale markets started. In most other cases [e.g., Argentina, Australia, the European Union, and the United Kingdom (UK)] there has been a progressive implementation of conditions defining qualified customers, starting with the largest customers under a multiyear phase-in transition program. (Note: although we realize that UK refers to the United Kingdom of Great Britain and Northern Ireland, occasionally, this book refers to England and Wales as the UK.)

A good indicator of competition and market maturity is the number of effectively *nonregulated* customers and total energy consumed outside regulated tariffs. For example, in California all customers were qualified in 1998, but two years later most of them continued under regulated tariffs, frozen at 10% below 1996 rates (not including charges to cover stranded costs). Later, during the electricity crisis in California of 2000 and 2001, retail choice was suspended. In Spain, on the other hand, the regulator adopted specific measures, such as the reduction of access tariffs, to promote the exit of regulated customers. At the end of 1999, of the more than 10,000 qualified customers about 80% were nonregulated customers, but the corresponding consumed energy was a small portion (2%) of the total consumption in Spain.

1.6. CONCLUSIONS

Restructuring and deregulation of the electricity industry is a movement with the aim of achieving lower prices to customers through cost savings. However, the brief history of this process shows that there is still much to be learned. Despite this, there is a consensus (1) to introduce competition into wholesale and retail markets by deregulating generation and opening retail and (2) continuing to regulate network activities.

But the experience also shows that those governments that started deregulation are continually revising their regulations. Argentina, California, England and Wales, and Spain, are still carrying out important revisions. The regulatory solutions adopted and the design of a transitional period to implement the new organizational structures are strongly influenced by the starting point of the industry and the political and institutional constraints in each country. To understand this continuous revision, this book explains the economic and regulatory principles behind electricity restructuring and focuses on some of the most representative experiences to illustrate its complexities.

CHAPTER 2

ELECTRICITY ECONOMICS

2.1. WHAT IS A MARKET?

Economics focuses on optimization and equilibrium. Macroeconomics addresses the general economy and asks whether macroeconomic indicators, such as inflation, unemployment, and the cost of capital, are in equilibrium. Microeconomics addresses the optimizing behavior of consumers and producers and asks whether the observed equilibrium prices and quantities in each market are *economically efficient* (defined below). There is no guarantee that freely operating markets will necessarily lead to efficiency. This chapter discusses the microeconomics of the electricity market. We will see that characteristics of the electricity market can lead to nonefficient prices and quantities. We will learn how to measure the social losses associated with both nonefficiency and the imposition of regulation.

2.1.1. Competitive versus Noncompetitive Markets

Buying and selling electricity involves at least three productive activities: *generation, transmission,* and *distribution,* including wholesale and retail distribution. (Other electricity industry services, e.g., ancillary services, are discussed in Chapter 5.) These activities can be bundled in a single market in which producers generate, transmit, and distribute electricity to consumers, or these activities can be unbundled. For more on electricity markets, see Stoft (2002, pp. 17–29).

When these activities are bundled, there is a single electricity provider. This provider can be owned by a national, regional, or local government, owned by a cooperative of consumers, or owned privately. Because of the ***monopoly*** (single seller) status of the provider, this firm can charge a price *above* the cost of production (including the cost of capital; see Chapter 3). To achieve economic efficiency, the ability to charge a price above the cost of production (known as *market power*) must be mitigated through some form of regulation (see Chapter 4) or through introducing competition (see Chapter 5).

Although we will assume that electricity markets are characterized by either *monopoly* or *competition*, there is a continuum of market types that depend on the number of sellers and their interaction. Under *perfect competition* the interaction of many buyers and sellers yields a market price equal to the cost of producing the last unit sold. This is the *economically efficient* solution.

Under monopoly, a single seller can reduce quantity, driving the price above the cost of production. Traditionally, the electricity industry has been dominated by national or local monopolies under price regulation to encourage economically efficient behavior. Under regulation, the regulator sets prices. As electricity generation is restructured, the role of the regulator becomes one of setting market guidelines to yield competitive conditions under which prices and quantities are similar to what they might be under the ideal of perfect competition. This chapter describes the characteristics of competition and monopoly in the electricity industry. Chapters 4 and 5 describe how to structure electricity markets so effective competition can set prices where appropriate and regulation can encourage efficiency where effective competition is not possible.

2.1.2. The Market Mechanism

The simplest method of describing the interaction between buyers and sellers in a market is to assume the existence of an auction. A hypothetical auctioneer (1) announces a range of prices to both buyers and sellers, asking market participants to reveal the quantity they are willing to buy or sell at *each* price and (2) determines a price that equates the quantity demanded by buyers with the quantity supplied by sellers. (Of course, most markets operate without an explicit auctioneer, however in many unbundled electricity systems, the Independent System Operator acts as the auctioneer; see Chapter 5.)

The quantity demanded at each price is what economists call "the demand schedule," or simply, ***demand***. (Note: here, demand is not the same as the common usage in electrical engineering, where it refers to the instantaneous capacity required by the load.) The quantity supplied at each price is the supply schedule,

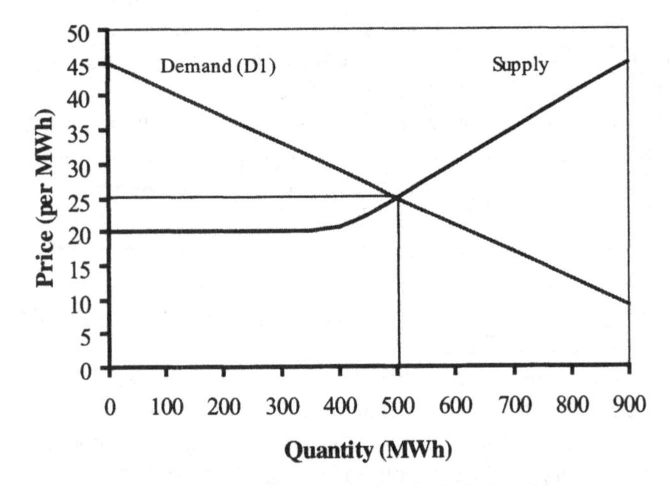

Figure 2.1. Supply and demand of electricity.

or simply, ***supply***. We represent supply and demand graphically in Figure 2.1. We represent price on the vertical ("y") axis and quantity on the horizontal ("x") axis. (Generally, we express quantity as a function of price, following standard practice in economics.) For example, to represent the market for electric power, the vertical axis represents the *price per megawatt-hour* (*MWh*) and the horizontal axis represents the *MWh quantity*. (Throughout this book we use both price per MWh and price per kWh. When expressed in US dollars, prices are in dollars per MWh, and mills per kWh. So U.S.$30/MWh equals 30 mills/kWh, where 1000 mills = 1 US dollar.)

In nearly all markets we find that as price falls, the *quantity demanded* rises. Buyers wish to purchase more at a lower price and less at a higher price. Therefore, the ***demand curve*** (the graphical representation of the demand schedule) has a negative slope. In most markets, we find that as price falls, the *quantity supplied* declines. At lower prices, suppliers are less willing or capable of producing. Therefore, the ***supply curve*** (the graphical representation of the supply schedule) has a nonnegative (i.e., zero or positive) slope. (The slope of the supply curve can be zero if suppliers are able, because of their costs, to supply more output at the same cost.)

We graphically represent the traded market price and quantity (traded during a period, such as during an hour, day, week, month, or year) by the intersection of the supply and demand curves. However, in many markets we usually only observe the intersection of supply and demand. On the one hand, if we assume that the demand schedule is fixed (i.e., the relationship between price and the quantity demanded is fixed), then shifts in supply reveal a series of intersections that trace out the demand curve. On the other hand, if we assume that the supply schedule is fixed (i.e., the relationship between price and the quantity supplied is fixed), then shifts in demand reveal a series of intersections that trace out the supply curve.

We make a distinction between changes in the quantity demanded (with changes in price) and shifts in demand. The demand for electricity can shift for different hours during the day or shift from one season to the next. Also, the supply curve can shift because of changes in cost. (In regulated markets firms reveal their supply curves to regulators, but there is no immediate mechanism that reveals demand, other than quantities consumed under regulated prices.)

The market price and quantity are in *equilibrium* during a period if buyers and sellers are satisfied with the market outcome. But suppose there is a shift in demand, for example from demand curve 1 (D_1) to D_2 in Figure 2.2. What will happen to the market price and quantity?

For example, assume an unusually cold, dark winter day. At all prices, more electricity will be demanded and the demand curve shifts to the right. At the original market price, P_1, buyers will want to purchase much more electricity than previously, Q_3. However, suppliers are unable to produce Q_3 at P_1. Instead, price rises, lowering the quantity demanded, and a new equilibrium is established at (Q_2, P_2). (See Exercise 2.2 on how this equilibrium is established.)

Throughout this discussion we have focused on a single market. These single markets are the focus of *microeconomics*. In reality, many markets interact. For ex-

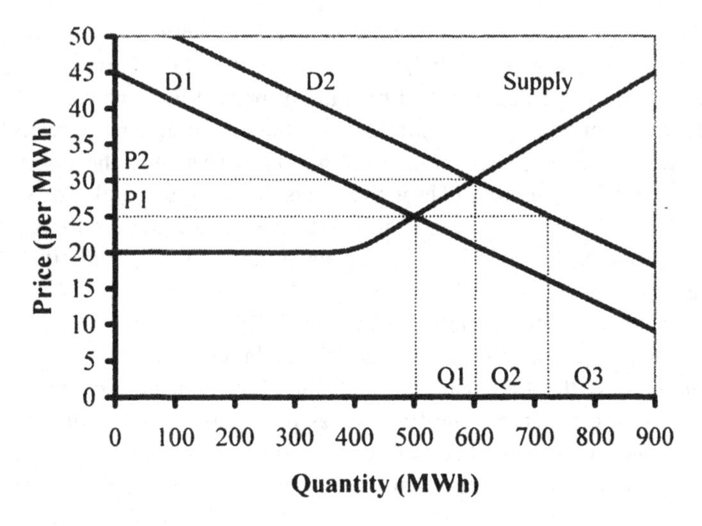

Figure 2.2. A shift in demand.

ample, the market for electricity and natural gas are intimately connected because electricity can be generated with natural gas, and electricity and natural gas are substitutes for direct heating applications. Further, markets in one geographic area can influence those in another area, or consumers can shift demand from one period to another. As we begin to consider the interaction between markets, we move from the analysis of *partial equilibrium* to *general equilibrium* for the economy as a whole. This general equilibrium is the focus of *macroeconomics*.

One macroeconomic influence that always complicates the analysis of markets is a change in the general level of prices. Although the general level of prices can fall, more often it rises. This is ***inflation*** (i). Because prices can inflate, to simplify the analysis, we will assume a constant price level. We define these as ***real prices***. To move from observed, or ***nominal prices***, to real prices, we must know the general level of inflation. We do this by comparing the total price of a set of goods and services over time. Changes in the total price of this constant set represent inflation. In the U.S., many payments are tied to the ***Consumer Price Index***, or CPI. Similarly, in the UK the ***Retail Price Index***, or RPI, is used. *Unless otherwise stated, throughout this book we discuss real prices (in terms of the purchasing power of a unit of currency at a particular time, e.g., 1999 U.S. dollars).*

2.1.3. Elasticity

Because the quantity demanded (Q_d) and the quantity supplied (Q_s) both change with changes in price, all market participants (including regulators) are better off knowing how these quantities change if price changes. The responsiveness of quantity to changes in price is ***price elasticity***. Because regulators set prices to achieve a total revenue target, they must know how market quantities respond to changes in price. If the price is set too high or too low, revenues will be lower than expected. It is a mistake to assume that by raising prices, total revenue will necessarily increase. Total revenue depends on price elasticity.

The responsiveness of changes in Q_d to price is the ***demand elasticity*** (E_d) and the responsiveness of changes in Q_s to price is the ***supply elasticity*** (E_s). We measure elasticity as the percentage change in quantity divided by a small percentage change in price. For example, if quantity decreases by 10% when price increases by 5%, then the elasticity is equal to –2. Because Q_d is negatively related to price (when price *increases*, demand *decreases*), demand elasticity is *negative*. (Because it is negative, we sometimes drop the negative sign; however, it is always implied.) To be more specific,

$$E_d = \%\Delta Q_d / \%\Delta P = (\Delta Q_d / Q_d)/(\Delta P/P)$$
$$= (\Delta Q_d/\Delta P) \cdot (P/Q_d) \tag{2.1}$$

where Δ represents a small incremental change in quantity or price *between two points* on the demand curve (see Exercise 2.1). Demand elasticity can also be expressed more precisely in its continuous form: $E_d = (\mathrm{d}\, Q_d/\mathrm{d}\, P) \cdot (P/Q_d)$, where d is an instantaneous change at *a single point* on the demand curve. If Q_d does not respond to price (the same quantity is purchased whether there are *small* increases or decreases in price), then the demand elasticity is zero, or *completely inelastic* at Q_d. If Q_d goes to zero with any increase in price, then the demand elasticity is (negatively) infinite, or *completely elastic*.

On the other hand, because Q_s increases with increases in price, the *supply elasticity* is normally positive. For example, if the regulator increases the price of electricity, it is likely that producers will be able to produce more electricity. If there is little change in Q_s, then supply is inelastic. If there is a great change in Q_s, then supply is elastic. Of course, both supply and demand elasticities depend on the initial price. At a high equilibrium price, demand might be elastic and supply might be inelastic, because (1) small decreases in price might induce more consumers to enter the market, but (2) producers might be producing all they can without investing in new capacity, so there is little change in supply. However, at a low equilibrium price demand might be inelastic and supply might be elastic, because small decreases in price will not induce much more consumption, but could lead to a large supply response.

Also, responsiveness changes over time. As consumers adjust to higher prices, their responsiveness can be more elastic. A sudden increase in the price of electricity could leave consumers with few alternatives. But with time, they could switch, for example, to more efficient electrical appliances. Further, a sudden increase in the price of electricity could lead to little supply response by electricity generators, but with time, new investments could be made in generating capacity. Another reason for a change in responsiveness over time is income change. As incomes rise, consumers change consumption patterns. This type of responsiveness is called ***in-***

come elasticity (equal to the percentage change in demand associated with a 1% change in income).

Also, changes in the price of substitutes (for example, natural gas is a substitute for electricity in some applications) or complements (for example, appliances that use electricity) can change demand for electricity. This type of responsiveness is called ***cross-price elasticity***. The more responsive demand is to changes in other prices, the more cross-price elastic it is. Increases in the price of *substitutes* generally lead to increases in demand for electricity, so the cross-price elasticity is positive. If the price of natural gas rises, consumers would prefer to consume relatively cheaper electricity. Here, electricity and natural gas are substitutes. *Decreases* in the price of *complements* can lead to *increases* in demand for electricity, so the cross-price elasticity is *negative*. If the prices of electric cars drop, more consumers buy more electric cars, and the demand for electricity would increase. (So, electricity and electric cars are complements.)

Because of the importance of understanding how changes in price might change demand or supply, microeconomics investigates the underlying structure of supply and demand. Because of the importance of understanding *producer* behavior under regulation, the remainder of this chapter will focus on the structure of *supply*. Although we will discuss demand in terms of the social costs of monopoly, interested readers should see Pindyck and Rubinfeld (2001, Chapters 3 and 4) for a more complete discussion of the structure of demand. (Also, Exercise 3.2 discusses the theory behind the economics of demand.)

2.2. COST AND SUPPLY

2.2.1. Economic Cost versus Accounting Costs

Before we discuss supplier (producer) behavior we must define cost. The economists' definition of cost is different from the accountants' definition (which usually refers to the price of inputs and outputs found on ***balance sheets***, which list the firm's assets, liabilities, and equities, giving the firm's financial position). Following the discussion above regarding real prices, economists define the cost of a scarce input relative to its alternative uses. Although we generally refer to the cost of an input in currency (e.g., dollars), implicit in measuring the cost of an input is its alternative use. This is the ***opportunity cost*** approach. It assumes that there are complete markets for all inputs. For example, the opportunity cost of an hour of labor is the best use of that hour in another activity. The opportunity cost of working another hour at the office might be spending another hour with your family. When allocating scarce resources to productive activities, efficient producers consider the best use of each input.

Generally, microeconomics focuses on two basic resources: *labor* (L) and *capital* (K). There are other inputs in production, such as *fuel* (F), but to keep the discussion simple we will focus on L and K and one output, represented by quantity, Q (for example, MWh of electricity).

What is the price of acquiring units of labor and capital? Although there are many types of labor, we will refer to the price of an average hour of labor in money *wages* (w). Implicitly, we assume that labor is allocated to its most productive use (the use with the highest opportunity cost) and that wages are determined in a competitive labor market. (However, in some locations a single employer can depress wages by exercising market power as a single *buyer*; this is *monopsony,* as opposed to *monopoly,* in which there is a single *seller.*)

The wage bill is the sum of wages paid to all forms of labor employed by a firm (or producer) and the average wage rate is the wage bill divided by the total number of hours of labor. (Generally, a firm is a productive entity, whether employing one person or a million people, or owned privately or publicly.)

Measuring units of capital and the price of capital is more complex. Capital can be measured in real terms as the number of machines, or in capacity terms, such as the generating capacity of an electric power plant. Or capital can be denominated in money terms at acquired prices. Because of the opportunity cost approach, we implicitly assume that physical capital is allocated to its best use and that all physical capital can be rented in a well-developed capital market. Therefore, if a piece of capital has a higher use elsewhere, its owner will *rent* it through the capital market to the producer who values it the most. Therefore, we refer to the price of capital charged to the user as a *rental rate* (r). (In Chapter 3 we refer to r as the *rate of return* to the provider of the capital and we refer to the rate charged on debt as the *rate of interest.*) If we measure capital by its cost of acquisition, then the rental rate is the rate that could be earned on money in alternative uses. If this money has been acquired from a bank, the rental rate on capital would be equal to the bank's *interest rate* on debt.

Also, another important economic concept is that of ***sunk cost***. Sunk costs are those costs that cannot be recovered when production ceases (Pindyck and Rubinfeld, 2001, Chapter 7). For example, many costs in *preparing* a site for the installation of new generating capacity are sunk. In fact, if it is uneconomic to move generating equipment to a new site, the *entire* cost of the generating capacity could be sunk. The importance of recognizing sunk costs is that they should not be taken into account when making decisions. If an old generating unit is uneconomic, then its cost of construction is sunk and should not be considered in determining whether to retire the unit; the firm should only consider the plant's operating costs.

2.2.2. Total, Average, and Marginal Costs

Associated with the basic inputs of labor and capital are the notions of variable and fixed costs. ***Fixed costs*** are fixed during some period. Although a cost might be fixed during a short period, such as a month, it could vary during a longer period, such as a year. We define the ***short run*** as a period during which there are *some* fixed costs. Further, we can define the following terms whereby technology describes how L and K are combined to produce Q:

Time	Cost	Technology
very short run	**all** costs are fixed	fixed
short run	**some** costs are fixed	fixed
long run	**no** costs are fixed	fixed
very long run	**no** costs are fixed	**not fixed**

Also, variable costs are those costs that vary in the short run with changes in output. Although some forms of labor, once hired, are fixed in the short run and some forms of capital are rented under agreements that depend on output, we will assume that labor is variable and capital is fixed in the short run.

Total Cost (*TC*) is the sum of ***Variable Cost*** (*VC*, e.g., the wage bill) and ***Fixed Cost*** (*FC*, e.g., the cost of renting capital). We represent this as

$$TC = w \cdot L + r \cdot K = VC + FC \tag{2.2}$$

Average cost (*AC*) is total cost divided by the quantity produced (*Q*):

$$AC = TC/Q = VC/Q + FC/Q = AVC + AFC \tag{2.3}$$

where
VC/Q is *Average Variable Cost* (*AVC*)
FC/Q is *Average Fixed Cost* (*AFC*)

An example of these cost terms can be found in Table 2.1 and in Figures 2.3a and 2.3b. Also, ***Marginal Cost*** (*MC*) is equal to the change in total cost with a unit change in quantity, *Q* (discussed below).

In determining how to allocate variable inputs, it is important to know the cost of producing a particular unit of output. The cost of producing a *particular* unit is its *marginal cost*. The marginal cost of producing the first unit includes *all* of the fixed costs and *some* of the variable costs. Therefore, marginal cost can be different in the

Table 2.1. Total, Average, and Marginal Cost

Q	Total Cost	Fixed Cost	Variable Cost	Average Cost	Average Fixed Cost	Average Variable Cost	Marginal Cost
100	7,250	5,000	2,250	73	50	23	25
200	10,000	5,000	5,000	50	25	25	30
300	13,250	5,000	8,250	45	17	28	35
400	17,000	5,000	12,000	43	13	30	40
500	21,250	5,000	16,250	43	10	33	45
600	26,000	5,000	21,000	43	8	35	50
700	31,250	5,000	26,250	45	7	38	55
800	37,000	5,000	32,000	46	6	40	60
900	43,250	5,000	38,250	49	6	43	65

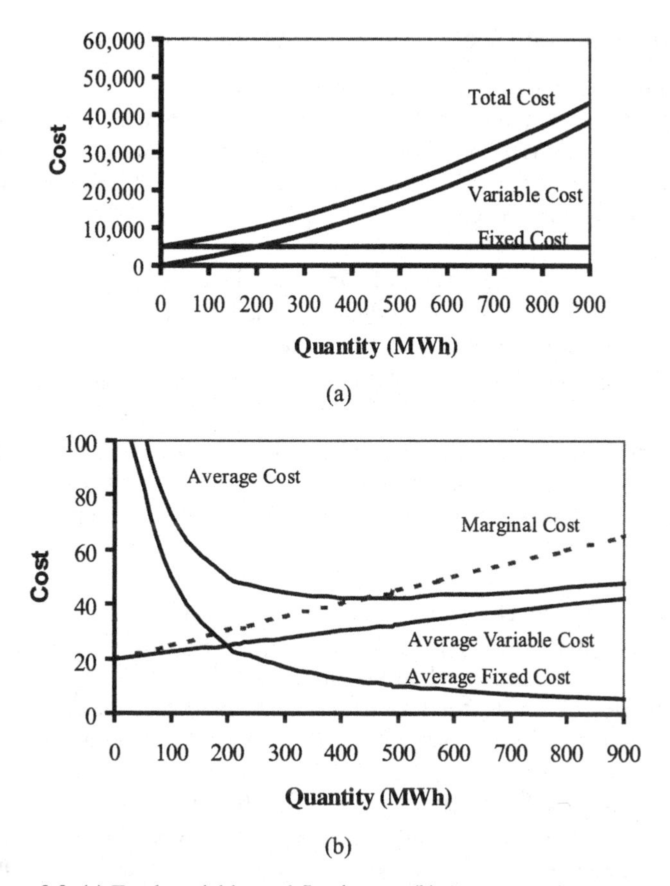

Figure 2.3. (a) Total, variable, and fixed costs. (b) Average and marginal costs.

short run, where some costs are fixed (i.e., *short-run marginal cost*), and in the *long run,* where no costs are fixed (i.e., *long-run marginal cost*).

If average variable cost is constant, marginal cost is level. If productive capacity becomes constrained, marginal cost rises. For example, the cost of producing a MWh changes as more MWhs are produced. Marginal cost can decrease as more electricity is produced, it can be level until capacity is constrained, and it can become very high when full capacity is reached (and diesel generators are started). Although marginal cost is associated with units of output, we will assume continuous changes in cost such that marginal cost (MC) can be represented as the first derivative of total cost with respect to quantity:

$$MC = \mathrm{d}TC/\mathrm{d}Q \tag{2.4}$$

where d is an instantaneous change in the variable. We will now use these cost concepts to discuss how cost changes with changes in the scale and the scope of operations.

2.2.3. Economies and Diseconomies of Scale and Scope

For many productive activities average cost is (1) high for low output, (2) lower in some range of output, and (3) high for high levels of output. This yields a familiar "U-shaped" average cost curve. This shape is a function of quantity: for small quantities, fixed costs are high; for large quantities, variable costs increase as quantity approaches production capacity. See Exercise 2.3.

1. Further, when average cost is falling, marginal cost is below average cost. This is because if a unit of production costs less than the average cost of all previous production, then average cost falls.
2. On the other hand, if a unit of production costs more than the average, then average cost rises. Then marginal cost is greater than average cost.
3. If average cost is constant, the cost of producing the next unit of production is equal to the average cost of producing the previous units of production, so marginal cost is equal to average cost, and the cost curves intersect, as in Figure 2.3b.

When a single-product firm experiences falling average cost with increases in output (and marginal cost is below average cost), ***economies of scale*** result. When average cost is equal to marginal cost (so average cost is neither rising nor falling), the firm experiences constant returns to scale. When a firm experiences increasing average cost with increases in output (and marginal cost is above average cost), diseconomies of scale result.

For some productive activities, fixed cost is a high proportion of total cost. For example, the construction of an oil well represents most of the cost of producing crude oil at a particular site. The incremental costs of pumping another barrel of oil are small (until most of the oil has been drained). Therefore, at a single production site, there are economies of scale in the production of oil.

Also, in the production of telephone service, most of the total cost consists of the fixed costs of constructing the communications network. The cost of placing a call (i.e., using the mechanisms for switching connections so people can communicate) is low. Therefore, the telecommunications network exhibits economies of scale. This is true of most network services, including transportation and electricity transmission and distribution. Further, for some generation technologies the cost of capital dominates the variable costs of production (consider hydroelectric and nuclear power). Therefore, an integrated electricity industry exhibits economies of scale, although some generation technologies (for which the cost of fuel is substantial) exhibit more usual-shaped cost curves.

In industries with economies of scale, average cost decreases with additional output (see Figure 2.4). So the firm with the largest output can produce at the lowest cost, driving competitors from the market. This is a ***natural monopoly*** situation. It is "natural" because of (1) the underlying characteristics of the production process and (2) the size of the market (see Exercise 2.3 and Berg and Tschirhart, 1988). Once competitors have been eliminated, a "natural" monopolist can exploit its position to drive prices above the cost of production. Because of these underlying char-

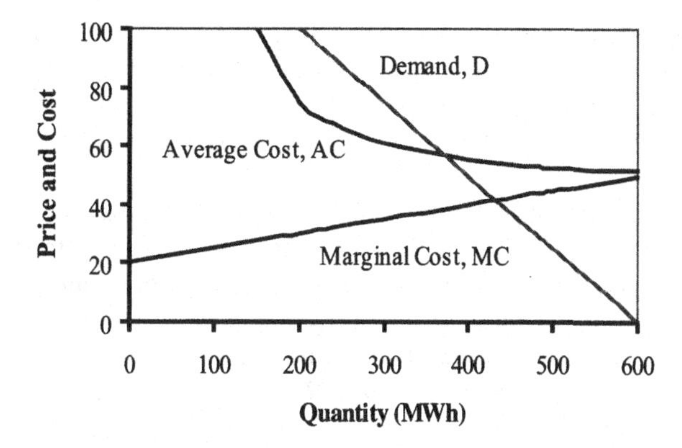

Figure 2.4. Demand and cost for the natural monopolist.

acteristics, a competitive market in a natural monopoly industry leads to the exploitation of market power. This is a case of ***market failure***: a competitive market fails to yield a market price equal to the marginal cost of production. In natural monopoly situations, a single firm can produce at the lowest total cost. However, because of market power, regulation (in some form) is usually required to yield prices closer to marginal cost.

Technical efficiency can arise in the joint production of two or more products. If the cost of producing two products by one firm is less than the cost of producing the same two products by two firms, the production process exhibits *economies of scope*. For example, if the joint production of electricity and heat can be done more cheaply than the production of electricity and heat separately, then there are economies of scope in this joint production process. Although there is not necessarily a functional relationship between economies of scale and economies of scope, both are important when considering whether a multiple product industry could lead to natural monopoly.

2.3. PROFIT MAXIMIZATION

2.3.1. What is Profit?

Before we go further, we must distinguish between *accounting profit* and *economic profit*. Accounting profits are the difference between revenues and accounting costs, i.e., those costs associated with an entry in the firm's accounts. Because economists use the opportunity cost of the inputs, there is not necessarily an entry in the accounts for every cost.

For example, if the firm borrowed money to finance the purchase of equipment and must pay the bank an amount each month on its loan, there would be an accounting entry for the cost of the loan. But if the firm's owner contributed savings to the firm, there would not necessarily be an entry in the accounts to pay the firm's owner for this contribution. Instead, the owner claims the accounting profit of the firm.

However, we should consider the opportunity cost of the owner's contribution. The owner could have lent those funds elsewhere. Therefore, *economic profits* are *above* the opportunity cost of capital, i.e., the rental rate on capital (r), as in Equation (2.2). Therefore, profit includes the opportunity cost of capital as a cost of the capital input.

Our primary assumption is that producers attempt to maximize the difference between revenues from their sales and the cost of production. This difference is ***profit***. Also, we assume that firms are *profit maximizers*. This is not to imply that all firms producing all goods and services are all profit maximizers, but profit maximization is a good first-order approximation of how firms behave.

To be more concrete, let ***total revenues*** (TR) equal the market price (P) of the product times the quantity (Q_j) sold: $TR_j = P \cdot Q_j$, where j designates the jth firm. Here, price is a function of quantity because of consumer demand response, so $TR_j = P(Q_j) \cdot Q_j$. Profit (PR_j) is the difference between TR_j and TC_j (from Equation 2.2):

$$PR_j = TR_j - TC_j \tag{2.5}$$

A profit-maximizing firm seeks to maximize *total* profit (not rate of return). If marginal cost increases with additional units, the firm should increase production as long as the marginal cost of producing one extra unit is less than the ***marginal revenue*** of selling the extra unit. For a competitive firm (which has no influence on the market price, so $\mathrm{d}\,P/\mathrm{d}\,Q_j = 0$), this means that it will produce at a quantity where marginal cost equals the market price. This quantity is found by maximizing PR with respect to Q_j and setting the result equal to zero:

$$\begin{aligned} \mathrm{d}PR_j/\mathrm{d}\,Q_j &= \mathrm{d}TR_j/\mathrm{d}Q_j - \mathrm{d}TC_j/\mathrm{d}\,Q_j \\ &= MR_j - MC_j = 0 \end{aligned} \tag{2.6}$$

At the maximum, MC (marginal cost) equals MR (marginal revenue). Further, because marginal revenue is a function of Q,

$$\begin{aligned} MR_j = \mathrm{d}TR_j/\mathrm{d}Q_j &= \mathrm{d}[P(Q_j) \cdot Q_j]/\mathrm{d}Q_j \\ &= (\mathrm{d}P/\mathrm{d}Q_j) \cdot Q_j + P \end{aligned} \tag{2.7}$$

by the chain rule of differentiation. Under competitive conditions, MR_j is equal to the market price, P, because $(\mathrm{d}P/\mathrm{d}Q_j) = 0$ (because changes in *individual* firm output are not large enough to influence the market price) in Equation (2.7). Therefore, the profit-maximizing firm chooses output such that price is just equal to marginal cost (substituting P for MR_j in Equation 2.6): $MC_j = P$. As discussed below, this is the *economically efficient* price.

When there are economic profits and *no barriers to entering* the market, several things happen under competitive market conditions:

1. Other firms enter
2. These entering firms increase output in the market

3. Prices decline (because of the outward shift in supply)
4. Economic profits decline

When there are economic losses and *no barriers to exiting* the market:

1. Firms exit
2. Output declines
3. Prices rise (because of an inward shift in supply)
4. Economic losses decline

With *no barriers to entry or exit*, we expect to see *economic profits* equal to zero in the long run, i.e., firms earn normal rates of return.

Considering the model of the market as an auction, as the market price rises, profit-maximizing firms adjust their output so that marginal cost is just equal to the market price. If average cost is above marginal cost (as it would be with the *left side* of the U-shaped average cost curve), then average cost would be above the market price. In this situation, the firm's total costs would be greater than its total revenues and it would be operating at a loss. If we assume that firms leave the market when they incur losses, firms would not produce under these conditions.

But in the natural monopoly case, marginal cost is below average cost in the *relevant range* of output. (For example, in Figure 2.4, $MC < AC$ when $Q < 600$ MWh.) If price were equal to marginal cost, then total revenues would be less than total cost and the natural monopolist would be operating at a loss. To remain in business, the natural monopolist must charge at least average cost. Therefore, the regulator must carefully choose a regulated price close to what it would be under efficient production (see Exercises 2.3–2.5).

On the other hand, if marginal cost is above average cost, total revenues would be above total costs and the firm would be earning a profit. Therefore, the individual firm's supply curve is its marginal cost curve above its average cost curve *in the long run* (see Figure 2.5). (*In the short run,* the firm's supply curve is its marginal cost curve above its average variable cost curve because in the short run its fixed costs have already been incurred. In Figure 2.5, the firm's supply curve is the entire marginal cost curve because $MC > AVC$ for $Q > 0$.) Finally, the *market supply curve* is simply the addition of all of the *firm's supply curves* at each price.

In summary, the theory of supply is based on

1. A production model of a firm's marginal cost
2. The assumption that firms choose input and output levels to maximize profit
3. The conclusion that firms enter and exit markets based on maximizing profit
4. The summation of each firm's supply curve to yield an industry supply curve

The theory of supply also yields a measure of industry returns known as the ***Producer Surplus*** (*PS*). Producer surplus is the difference between the market price and the variable cost of production, summed over output (see Figure 2.6). With a

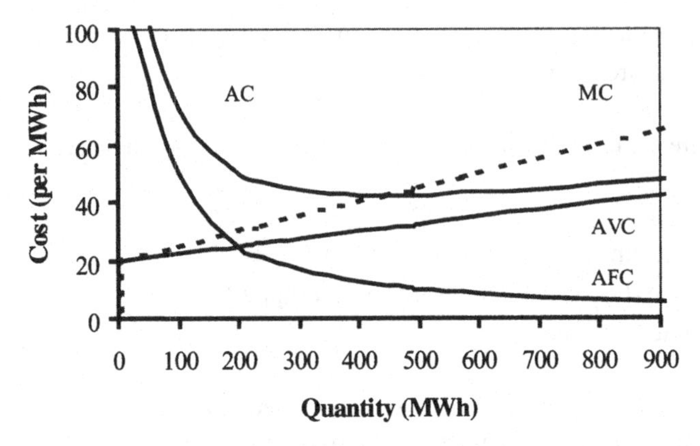

Figure 2.5. The firm's supply curve in the short run.

positively sloped supply curve, there are a variety of firms with different variable costs of production. But the lowest-cost firm receives the same price as the highest-cost firm. Therefore, a lower-cost firm earns a higher producer surplus than a higher-cost firm. The difference between total revenue and variable cost is producer surplus: $PS = TR - VC$ from Equation (2.2). So, in the short run the difference between producer surplus and profit is fixed cost: $PS - PR = FC$. However, in the long run producer surplus is equal to profit, because in the long run all costs are variable, so $TC = VC$ and $PS = TR - TC$.

For example, in a competitive electricity-generating market (as in California, see Chapter 6), there is one market price for each hour during the day. All producers receive the same price for a MWh of electricity. Generators with low variable costs of

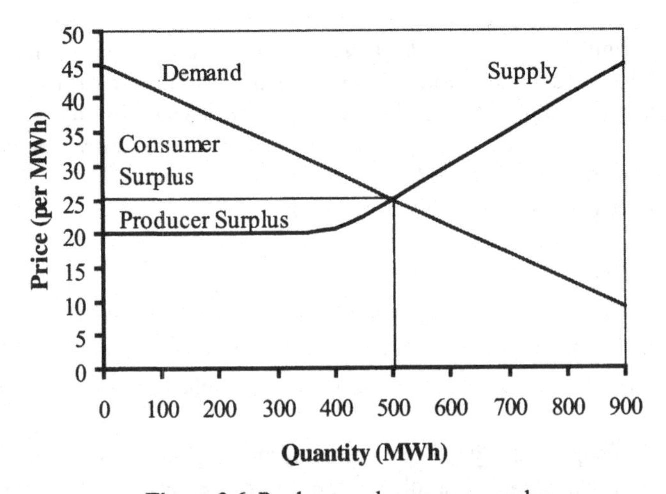

Figure 2.6. Producer and consumer surplus.

production (e.g., nuclear power plants) receive the same revenue as generators with high variable costs of production (e.g., oil-fired power plants). Low (variable) cost generators have a higher producer surplus than high (variable) cost generators, but *profits depend on both variable costs and fixed costs* of the two types of generators.

2.3.2. What is Economic Efficiency?

Above, we introduced the term "efficiency." Here, we clarify the difference between "technical efficiency" and "economic efficiency." ***Technical efficiency*** implies that the maximum output has been produced with a given set of inputs. This also implies that the minimum levels of inputs have been used to produce a given level of output. ***Economic efficiency*** implies that the maximum output has been produced at a given (opportunity) cost, or that a minimum (opportunity) cost has been achieved for a given level of output.

Under competition, minimum cost is achieved with price equal to marginal cost. This is the economically efficient price; it lets the consumer know the cost of producing another unit. When price does not reflect marginal cost, the market has *failed* to produce efficient prices.

For example, if the price of electricity is set below the efficient price (marginal cost), (1) consumers will demand more than the economically efficient level and (2) producers will produce less than the economically efficient level. If the price of electricity is set above marginal cost, (1) consumers will demand less than the economically efficient level and (2) producers will produce more than the economically efficient level. But if price is set at marginal cost for a natural monopolist, revenues will be too low to sustain production (see Exercise 2.4). Regulators must find a price that will encourage production at minimum cost (see Chapters 4 and 5).

2.4. Social Surplus: Consumer and Producer Surplus

Although we did not develop a complete theory of demand (Pindyck and Rubinfeld, 2001, Chapter 4), we now introduce an important idea that follows from the microeconomic theory of consumer behavior. Like optimizing firms, we assume that consumers purchase goods and services to maximize their well being. If they are able to purchase something at less than the maximum price they would be willing to pay, they enjoy a surplus equal to the difference between the market price and what they would have been willing to pay (as revealed to the auctioneer). This is ***Consumer Surplus*** (*CS*).

Assuming that all consumers pay the same price in the market, those consumers with a high willingness to pay enjoy a higher consumer surplus than those with a lower willingness to pay. Consumer surplus can be added across all consumers in a market. This total is consumer surplus (see Figure 2.6). For example, if electricity is sold to all consumers at the same price, consumers with access to alternative sources of energy, such as natural gas, will not place as high a value on electricity as those consumers who have no other alternatives.

Social Surplus is the sum of producer surplus and consumer surplus. It represents the surplus to society from the provision of goods or services. It is the difference between the total benefit to consumers minus the total cost of production. This surplus is divided between buyers and sellers as consumer surplus and producer surplus. Social surplus is greatest when the market price is equal to the marginal cost of producing the last unit sold (as represented by the supply curve).

For example, assume that in a regulated market the regulated tariff or price (P_r) is set above marginal cost (see Figure 2.7, where P_r is set such that $Q_d = AC$). What is the impact on producers and consumers? Producers supply less (Q) than they would if the tariff were P (in the short run). There is a decrease in consumer surplus because less is produced. Also, the decrease in producer surplus is not entirely transferred to consumers. Some is lost completely. These losses are known as ***deadweight losses***: society as a whole is worse off because there are social losses from reduced production.

Regulated tariffs are not the only source of deadweight loss. When the long-run market price does not reflect the cost of production, there is market failure and a loss of social surplus. For example, in some markets the costs of production are not well defined. This includes markets in which there are external effects (***externalities***), such as pollution. If the cost of coal does not include the cost of pollutants, the price of coal is socially too low. Coal consumers purchase more coal than is socially optimal because they do not pay the costs of pollution. Too much coal is produced and those affected by the pollutants are forced to pay for the pollution directly or indirectly. Another source of social surplus loss occurs under monopolization, which we discuss next.

2.5. MARKET POWER AND MONOPOLY

2.5.1. Maximizing Profit under Monopoly

We consider the efficiency problems caused by monopoly in the final section of this chapter. We examine the profit-maximizing behavior of the monopolist without considering the origins of the monopoly. The monopoly could have arisen as the outcome of competition in a natural monopoly situation, it could have developed from an innovation protected by patent rights, or it could have formed through the collusion of all firms in an industry. See Stoft (2002, Part 4).

Because the monopolist is a profit maximizer, it follows the same output selection rule as profit maximizers in competitive industries (see Figure 2.8):

$$MR = (\mathrm{d}P/\mathrm{d}Q_j) \cdot Q_j + P = MC \tag{2.8}$$

1. If the monopolist produces less than the profit-maximizing quantity, it loses profit by producing too little and charging a price that is too high.
2. If the monopolist produces more than the profit-maximizing quantity, it loses profit by producing too much and charging a price that is too low.

Like the competitor, the monopolist chooses quantity to maximize profit where

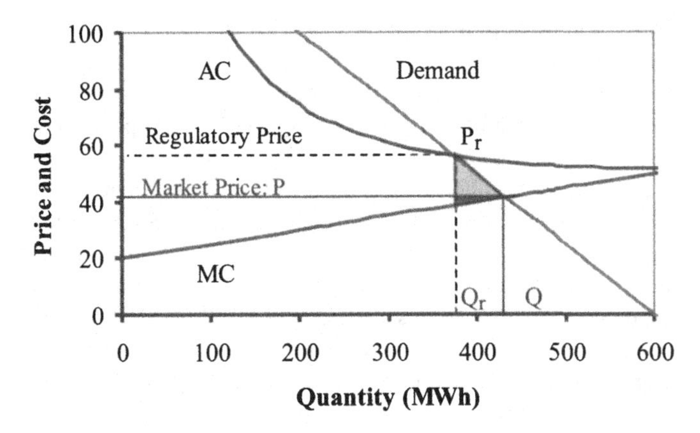

Figure 2.7. Deadweight losses from regulation.

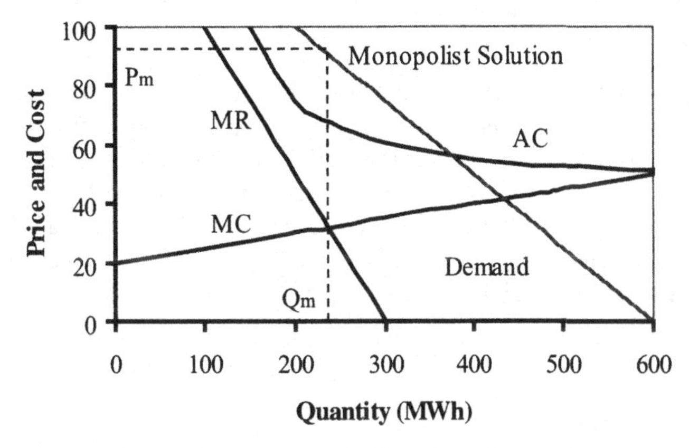

Figure 2.8. The monopolist's profit-maximizing output choice.

$MR = MC$. Unlike the competitor, the monopolist has the power to influence price. It chooses the price given by the demand curve. (If the monopolist can charge different prices to different consumers, it can transform more consumer surplus into profit. This is *price discrimination* and is discussed in Chapter 4.) Therefore, the monopolist sets (1) *quantity,* where marginal revenue is equal to marginal cost, and (2) *price* by "charging what the market can bear," i.e., by choosing the price given by demand. If the demand curve is not completely elastic, its price will be above marginal revenue, and thus above marginal cost.

2.5.2. Deadweight Loss from Monopoly Power

How great is the loss of social surplus from the exercise of monopoly (market) power? This can be determined by comparing the producer and consumer surplus under competitive and monopoly conditions (see Figure 2.9). Here, we see that by raising price above marginal cost, an amount equal to the area $Q_m \cdot (P_m - P_c)$ is transferred from consumers to producers. But there is also a *deadweight loss* equal to the sum of two triangular areas: (1) $\frac{1}{2} \cdot (Q_c - Q_m) \cdot (P_m - P_c)$ plus (2) $(Q_c - Q_m) \cdot P_c - \int MC$, where the last term is the integral of MC from Q_m to Q_c, i.e., the difference between total cost of producing Q_c and the total cost of producing Q_m (see Exercise 2.6). This deadweight loss is a loss of social surplus because the consumer does not enjoy it as consumer surplus and the monopolist does not earn it as profit. The size of the deadweight loss from monopoly pricing depends on demand elasticity. The monopolist is less able to raise prices if demand is more elastic, therefore there is less deadweight loss. Further, the monopolist might not minimize cost, spending resources to secure its monopoly power or leverage that power into other markets. This increases the social cost of monopoly.

2.5.3. Response to the Exercise of Monopoly Power: Regulation and Antitrust

On the other hand, the monopolist's power might not be secure. For example, if the monopolist is the only supplier of fuel oil in a region, increases in price above some level will induce competitors to enter the market. Therefore, if there are no barriers to entry and marginal cost is above average cost, competitive markets can be relied upon to discipline potential or temporary monopolists. When there are barriers to entry (either technical or legal), institutional mechanisms must be constructed to reduce losses in social surplus from the exercise of monopoly power. Two approaches can be used:

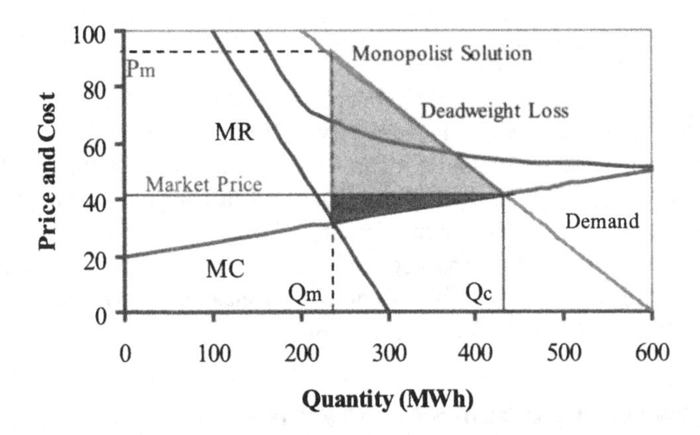

Figure 2.9. Deadweight loss under monopoly.

1. *Price or tariff regulation,* under which competitive markets are difficult to establish and sustain
2. *Antitrust enforcement* in established markets

Under regulation, the monopolist exchanges its control over price for legal rights. For example, the electric monopoly accepts the duty to serve all customers and submits to tariff (price) and investment regulation in exchange for protection from competitors and other legal rights, such as the right to purchase land for transmission lines. Although regulation focuses on tariffs that provide adequate revenues, regulatory authorities must also be concerned with

- Output levels
- The quality of output
- The construction of new capacity
- Attempts by regulated firms to extend their monopoly power into other markets

Price regulation requires regulated firms to provide so much technical and financial information to regulatory authorities that conflicts inevitably arise between regulators trying to maximize social surplus and monopolists trying to maximize profits. Therefore, there has been an attempt to redesign regulatory institutions to minimize information requirements while trying to mimic market efficiency. Chapter 4 reviews regulatory institutions in the electric utility industry. Chapter 5 reviews attempts to introduce market mechanisms into electric utility regulation, in particular into the generation and retail sale of electricity. Chapters 6–9 provide case studies of how market mechanisms have been recently introduced in the electric utility industry in several countries.

Another approach to reducing monopoly power is through antitrust enforcement. Although the explicit application of antitrust is usually not appropriate to regulated industries, as markets are introduced into electricity generation, antitrust principles must be applied to these deregulated markets. See, for example, FTC (1996) on merger policy in the electric utility industry. Antitrust regulations can be enforced in two ways:

1. Giving antitrust enforcement power to government institutions
2. Giving parties injured by the exercise of monopoly power the right to bring legal action against the monopolist and demand compensation for damages

A complete review of antitrust laws is beyond the scope of this chapter. Each country has devised laws and enforcement mechanisms to limit the formation of monopoly by either one firm or a group of firms acting together to raise prices, limit output, or impose barriers to entry. Further, there is great tension between the exercise of antitrust authority and the granting of legal monopolies to encourage some social benefit, such as the issuance of patents and other protections of intellectual property rights. Finally, when privatizing nationally owned, government monopolies, great care must be taken to avoid transferring monopoly power to private owners if regulatory and antitrust authorities are not well established.

EXERCISE 2.1. LINEAR AND LOGARITHMIC DEMAND FUNCTIONS

There are two popular representations of demand functions. These are the linear and the logarithmically demand functions:

$$P = a - b \cdot Q_d \tag{2.9a}$$

and

$$\ln P = c - d \cdot \ln Q_d \tag{2.10a}$$

Often, quantity demand, Q_d, is represented as a function of price, P, even though price and cost are *always* plotted on the vertical axis and quantity is *always* plotted on the horizontal axis. Solving for Q_d (or $\ln Q_d$) as a function of P (or $\ln P$),

$$Q_d = (a/b) - (1/b) \cdot P \tag{2.9b}$$

or

$$\ln Q_d = (c/d) - (1/d) \cdot \ln P \tag{2.10b}$$

Demand elasticity is the percentage change in quantity with respect to a small percentage change in price (e.g., 1%). First, with respect to discrete changes in price and quantity:

$$E_d = \%\Delta Q_d / \%\Delta P = (\Delta Q_d / \Delta P) \cdot (P/Q_d)$$

Second, with respect to continuous changes in price and quantity:

$$\begin{aligned} E_d &= (\mathrm{d}Q_d/\mathrm{d}P) \cdot (P/Q_d) = (\mathrm{d}Q_d/Q) \cdot (\mathrm{d}P/P) \\ &= \mathrm{d} \ln Q_d / \mathrm{d} \ln P \end{aligned}$$

For the linear example, if $E_d = -1$ (negative unitary elasticity), from Equation (2.9b)

$$(\Delta Q_d / \Delta P) = -(1/b)$$

and from Equation (2.9a)

$$P = a - b \cdot Q_d$$

Substituting these into the definition,

$$E_d = (\Delta Q_d/\Delta P) \cdot (P/Q_d)$$
$$= -(1/b) \cdot [(a - b \cdot Q_d)/Q_d] = -(a/b \cdot Q_d) + 1$$

Setting $E_d = -1$ and solving for Q_d,

$$Q_d = (1/2) \cdot (a/b)$$

i.e., a linear demand curve has (negative) unitary elasticity where quantity is one-half the horizontal intercept. Substituting this value for Q_d into Equation (2.9a) and solving for P,

$$P = (1/2) \cdot a$$

i.e., *demand elasticity* is *unitary* where price equals one-half the vertical intercept. For the logarithmic example, $E_d = \mathrm{d} \ln Q_d/\mathrm{d} \ln P$. From Equation (2.10b), $E_d = -(1/d)$. Therefore, if $E_d = -1$ (negative unitary elasticity), $-(1/d) = -1$ and $d = 1$. The demand elasticity of a logarithmic demand curve is *constant* and equal to $-(1/d)$. (Elasticity does not depend on c in Equations 2.10a and 2.10b, c shifts the demand equation vertically, but does not influence its shape.)

2.1.1. If $a = 45$ and $b = 0.04$ in Equation (2.9a), then

$$P = 45 - 0.04 \cdot Q_d \tag{2.9c}$$

Determine E_d for $Q_d = 400$, 600, and 800. Does demand elasticity become more elastic or more inelastic as quantity increases?

2.1.2. Determine the value of c so that Equation (2.10a) is tangent to Equation (2.9c) at $E_d = -1$. Graph these equations for price between 0 and 50 and quantity between 0 and 900.

2.1.3. If demand is actually logarithmic with unitary elasticity, determine the percentage error in predicting the quantity response when using a linear demand curve (Equation 2.9c) for $P = 30$, 25, 20, and 15.

EXERCISE 2.2. A SHIFT IN DEMAND AND A NEW EQUILIBRIUM PRICE (A COBWEB MODEL)

A linear demand equation is represented as

$$Q_d = a - b \cdot P \tag{2.11a}$$

and linear supply is represented as

$$Q_s = -c + d \cdot P \tag{2.12a}$$

with the equilibrium condition that $Q_d = Q_s$. However, there can be lags in supply because firms cannot respond immediately to changes in demand. Assume that there is a demand shift and the supply responds to last period's price (where a period can be an hour, day, week, month, or year). The new demand and supply system can be represented as

$$Q_{dt} = a' - b' \cdot P_t \tag{2.11b}$$

Supply is represented as a function of lagged price, P_{t-1}:

$$Q_{st} = -c + d \cdot P_{t-1} \tag{2.12b}$$

Substituting these into the equilibrium condition, $Q_{dt} = Q_{st}$:

$$a' - b' \cdot P_t = -c + d \cdot P_{t-1} \tag{2.13a}$$

or

$$b' \cdot P_t + d \cdot P_{t-1} = a' + c \tag{2.13b}$$

Shifting ahead one period and rearranging terms,

$$P_{t+1} + (d/b) \cdot P_t = (a' + c)/b' \tag{2.13c}$$

This is a first-order difference equation (a "discrete differential equation") that, when solved, yields the following time path for price:

$$P_t = (P_0 - P^*) \cdot (-d/b)^t + P^* \tag{2.14}$$

where P^* is the long-run equilibrium price given by

$$P^* = (a' + c)/(b' + d) \tag{2.15}$$

2.2.1. Show that the equilibrium solution to the following supply and demand system is $P_0 = 25$.

$$Q_d = 1125 - 25 \cdot P \tag{2.16a}$$

$$Q_s = 20 \cdot P \tag{2.16b}$$

2.2.2. Determine the price path and P^* for a shift in demand (D_2) to

$$Q_d = 1350 - 25 \cdot P \tag{2.17}$$

2.2.3. Calculate prices and excess demand for 6 periods. Plot the price path.

EXERCISE 2.3. RETURNS TO SCALE IN PRODUCTION AND COST

We define returns to scale in the context of production and cost functions. The *production function* is

$$Q = f(L, K) \tag{2.18a}$$

If all inputs are multiplied by a positive constant, m, then

$$Q \cdot m^g = f(m \cdot L, m \cdot K) \tag{2.18b}$$

where g measures the returns to scale. For example, if $m = 1.1$, then all inputs increase by 10%. m^g indicates whether output increases by 10%, less than 10%, or more than 10%:

1. If $g = 1$, then the production function exhibits constant returns to scale. If $m = 1.1$, output increases by 10%.
2. If $g > 1$, there are increasing returns to scale. If $m = 1.1$, output increases by more than 10%.
3. If $g < 1$, there are decreasing returns to scale. If $m = 1.1$, output increases by less than 10%.

A simple representation of the production function is the Cobb–Douglas model (Cobb and Douglas, 1928):

$$Q = c \cdot L^a \cdot K^b \tag{2.19a}$$

or

$$\ln Q = c + a \cdot \ln L + b \cdot \ln K \tag{2.19b}$$

(Note: the parameter c can be interpreted as a measure of technology, i.e., as c increases, the same level of L and K produces greater output.) If L and K both change by m, then

$$\begin{aligned} f(m \cdot L, m \cdot K) &= c \cdot (m \cdot L)^a \cdot (m \cdot K)^b \\ &= c \cdot (m^a \cdot L^a) \cdot (m^b \cdot K^b) \\ &= m^{a+b} \cdot c \cdot L^a \cdot K^b \\ &= m^{a+b} \cdot Q \end{aligned} \tag{2.20}$$

So $a + b$ measures the returns to scale:

1. If $a+b = 1$, there are constant returns to scale
2. If $a+b > 1$, there are increasing returns to scale
3. If $a+b < 1$, there are decreasing returns to scale

Similarly, the *cost function* is defined as the cost minimizing solution to the following constrained optimization problem (Varian, 1992, Chapter 1):

$$C(w, r, Q) = \min w \cdot L + r \cdot K$$

subject to

$$Q = f(L, K) \tag{2.21}$$

For the Cobb–Douglas production function one can solve for K (or L) in $Q = c \cdot L^a \cdot K^b$ and substitute:

$$\min w \cdot L + r \cdot c^{-1/b} \cdot Q^{1/b} \cdot L^{-a/b} \tag{2.22}$$

Minimizing this with respect to L and setting the result equal to 0 (ignoring the second-order conditions),

$$w - (a/b)\, r \cdot c^{-1/b} \cdot Q^{1/b} \cdot L^{-a+b/b} = 0 \tag{2.23}$$

Solving for L (the firm's demand equation for labor),

$$L = c^{-1/a+b} \cdot (a \cdot r/b \cdot w)^{\,b/a+b} \cdot Q^{1/a+b} \tag{2.24}$$

Solving for K in the same manner (the firm's demand equation for capital),

$$K = c^{-1/a+b} \cdot (a \cdot r/b \cdot w)^{-a/a+b} \cdot Q^{1/a+b} \tag{2.25}$$

Substituting these into the total cost equation,

$$C(w, r, Q) = d \cdot w^{a/a+b} \cdot r^{b/a+b} \cdot Q^{1/a+b} \tag{2.26a}$$

or

$$\ln C = \ln d + (a/a+b) \cdot \ln w + (b/a+b) \cdot \ln r + (1/a+b) \cdot \ln Q \tag{2.26b}$$

where

$$d = c^{-1/a+b} \cdot [(a/b)^{\,b/a+b} + (a/b)^{-a/a+b}]$$

This is the *Cobb–Douglas cost function*. If there are constant returns to scale in the

Cobb–Douglas production function, $(a + b = 1)$ and the Cobb–Douglas cost function becomes

$$TC(w, r, Q) = d \cdot w^{a} \cdot r^{1-a} \cdot Q \tag{2.27a}$$

or

$$\ln TC = \ln d + a \ln w + (1 - a) \ln r + \ln Q \tag{2.27b}$$

If there are *increasing* returns to scale (with cost-minimization), total cost increases *less* than the increase in the cost of inputs. For example, if all inputs increase by 10%, costs increase by less than 10%. If there are *decreasing* returns to scale, total costs increase more than the increase in the inputs. In this framework we can define average cost and marginal cost:

$$AC = TC/Q = d \cdot w^{a/a+b} \cdot r^{b/a+b} \cdot Q^{(1/a+b)-1} \tag{2.28}$$

$$MC = \mathrm{d}TC/\mathrm{d}Q = (1/a+b) \cdot d \cdot w^{a/a+b} \cdot r^{b/a+b} \cdot Q^{(1/a+b)-1}$$

$$MC = (1/a+b) \cdot AC \tag{2.29}$$

1. If $a+b = 1$, then $MC = AC$ with constant returns to scale.
2. If $a+b > 1$, then $MC < AC$ with constant returns to scale.
3. If $a+b < 1$, then $MC > AC$ with constant returns to scale.

In other words, *with continuously increasing returns to scale MC is always less than AC.* However, *few* production functions exhibit continuously increasing returns to scale. At some constraint, average costs begin to increase and marginal cost is greater than average cost. But if market demand is less than the quantity where $MC > AC$, a "natural monopoly" can develop.

Nerlove (1963) provides an early estimation of a Cobb–Douglas cost function for electric utilities. His final form was similar to Equation (2.27b). However, he introduced the price of fuel ($\ln f$) and dropped the price of capital ($\ln r$), because there was little variation in the measured price of capital across the 145 electric utilities that constituted his sample. Further, he found that total cost was a quadratic function of output (Q in MWh):

$$\ln TC = \beta_0 + \beta_1 \cdot \ln w + \beta_2 \cdot \ln f + \beta_3 \cdot \ln Q + \beta_4 \cdot (\ln Q)^2 \tag{2.30}$$

Here, $(\mathrm{d} \ln TC/\mathrm{d} \ln Q)$ equals $(\beta_3 + 2 \cdot \beta_4 \cdot \ln Q)$.

Building on Nerlove, Christensen and Greene (1976) define *scale economies* as

$$1 - \mathrm{d} \ln TC/\mathrm{d} \ln Q = 1 - (\beta_3 + 2 \cdot \beta_4 \cdot \ln Q) \tag{2.31}$$

This yields positive values for increasing returns to scale and negative values for decreasing returns to scale. Although Nerlove found economies of scale for electric utilities, increasing returns to scale were far less for larger utilities than smaller utilities. Christensen and Greene (1976), using a more general functional form (the translog), found that utilities in 1970 with outputs above 25 TWh annually did not exhibit increasing returns to scale.

Because of the importance of the relationship of TC and Q in determining returns to scale, many analyses of cost, output, and returns to scale use a linear reduced form of (2.30):

$$TC = FC + \beta_3 \cdot Q + \beta_4 \cdot Q^2 \tag{2.32a}$$

This is a quadratic *cost equation* or *cost curve* and can be generalized to the *cubic cost curve*:

$$TC = FC + \beta_3 \cdot Q + \beta_4 \cdot Q^2 + \beta_5 \cdot Q^3 \tag{2.32b}$$

Again, one minus the elasticity of cost with respect to output, $1 - [(\mathrm{d}TC/\mathrm{d}Q)/(TC/Q)]$, yields positive values for increasing returns to scale and negative values for decreasing returns to scale.

2.3.1. Assuming that fixed cost is \$5000 and $VC = 20 \cdot Q + 0.025 \cdot Q^2$, find total cost, fixed cost, variable cost, average fixed cost, average variable cost, average cost, and marginal cost for quantities 0, 100, 200, . . . , 900.

2.3.2. There are increasing returns to scale when AC is falling and decreasing returns to scale when AC is rising. For the cost equation identified above, determine the quantity at which AC begins to rise. Also determine the quantity at which $MC = AC$.

2.3.3. If fixed costs increase to \$10,000, at which quantity does $MC = AC$? Plot MC and AC for quantities between 0 and 600 and $AC < 100$.

2.3.4. The firm's supply curve is equal to MC above average variable cost. On a plot of MC, AVC, and AC show the firm's supply curve.

2.3.5. Assume the demand equation is $P = 150 - 0.25 \cdot Q$. Plot demand on the same graph as the firm's supply. Determine the quantity and price where demand equals supply. This is the competitive market solution.

EXERCISE 2.4. CALCULATING A REGULATED TARIFF

The primary problem with marginal cost pricing can be seen in Figure 2.7. Because marginal cost is less than average cost, the competitive market solution (i.e., where supply equals demand) results in losses for the firm, here assumed to be a monopolist. As we discuss in Chapter 4, there are several solutions to this problem. One solution is for the regulator to set a tariff so that average revenue is equal to average cost. Because $TR = P(Q) \cdot Q$, average revenue is simply the demand equation:

$$AR = TR/Q = P(Q) \tag{2.33}$$

i.e., price as a function of quantity demanded.

2.4.1. Under the cost (supply) and demand conditions in Exercise 2.3.5, determine the firm's average cost and total losses if the tariff is equal to marginal cost.

2.4.2. Under these same conditions, determine the price and quantity if the regulator sets a tariff so that Average Revenue, AR, is equal to Average Cost, AC.

EXERCISE 2.5. CALCULATING SOCIAL SURPLUS UNDER COMPETITION AND REGULATION

If consumers are able to purchase something at less than the maximum price they would be willing to pay, they enjoy a surplus equal to the difference between the market price and what they would have been willing to pay. *Consumer Surplus, CS,* is the integral of surplus for all consumers in the market. For example, consumer surplus at P_c and Q_c, the competitive equilibrium price and quantity, is

$$CS = \int [P(Q) - P_c]\, dQ \qquad \text{from } Q = 0 \text{ to } Q_c \tag{2.34}$$

The theory of supply also yields a measure of industry returns known as *Producer Surplus, PS*. The difference between Total Revenue, TR, and Variable Cost, VC, is producer surplus: $PS = TR - VC$ over all Q. Because VC at $Q = Q_c$ (at the competitive equilibrium quantity) is the integral of marginal cost from $Q = 0$ to Q_c:

$$\begin{aligned} PS &= TR - VC \\ &= P(Q) \cdot Q - \int MC \qquad \text{from } Q = 0 \text{ to } Q_c \end{aligned} \tag{2.35}$$

Social Surplus is equal to consumer surplus plus producer surplus. *Deadweight Loss* is the loss of social surplus when efficiency is not achieved.

2.5.1. Determine the consumer and producer surplus and the social surplus for the competitive solution defined in Exercise 2.3.5.

2.5.2. Determine the consumer and producer surplus and the social surplus for the regulated tariff in Exercise 2.4.2.

2.5.3. Determine the deadweight loss of imposing a regulated tariff equal to average cost in Exercise 2.4.2.

EXERCISE 2.6. CALCULATING DEADWEIGHT LOSS UNDER MONOPOLY

To maximize profits the monopolist attempts to choose quantity so that

$$MR = (\mathrm{d}P/\mathrm{d}Q) \cdot Q + P = MC$$

Given this quantity, the monopolist sets price given by demand. With linear demand, $P = a - b \cdot Q_{-b}$,

$$MR = b \cdot Q_j + (a - b \cdot Q)$$
$$= a - 2 \cdot b \cdot Q = MC$$

i.e., with linear demand, the marginal revenue curve has a slope twice that of the demand curve.

2.6.1. With the supply and demand conditions as in Exercise 2.3.5, determine the monopolist's profit-maximizing quantity and price. Calculate total revenues, total cost, and profit.

2.6.2. Calculate consumer and producer surplus under the monopolist's profit-maximizing quantity and price. Calculate deadweight loss and compare with the deadweight loss under regulation.

CHAPTER 3

THE COST OF CAPITAL

One of the most difficult tasks of an electric utility is to determine its cost of capital. As we saw in Chapter 2, total cost is equal to variable costs plus fixed costs. As we will see in Chapter 4, under regulation the regulator must determine (1) which operating expenses are appropriate and (2) whether customers are paying prices that ensure sufficient long-run security of supply.

Under regulation, a utility is allowed to recover reasonable costs incurred in the provision of service. Under deregulation, generators are not guaranteed cost recovery, but most transmitters and distributors remain regulated; see Chapters 4 and 5. Typically, the definition of reasonable costs is based on a review and evaluation of the accounting costs incurred. However, as discussed in Chapter 2, there are some costs that are not readily identifiable in this way, such as the cost of equity.

3.1. WHAT IS THE COST OF CAPITAL?

Financing the capital cost of new investment can involve borrowing funds from financial institutions that charge a ***rate of interest*** (the *rental rate* in Chapter 2). This is ***debt*** capital. Determining the appropriate cost of capital for these funds is easy. The rate of interest is the cost of capital charged by financial institutions. However, some financial capital is provided by the owners or shareholders of the firm or from profits earned by the firm. This is ***equity*** capital. The ***rate of return*** is the rate that the owners of the firm earn on their equity. When the firm is a regulated electric utility, the regulator must decide the appropriate rate of return to charge customers to remunerate the providers of equity capital. This chapter discusses the principles used to determine this rate of return.

To simplify our discussion, we begin by assuming that there is a single market for financial capital. In this market, firms announce how much they are willing to borrow at each rate of interest and individual investors announce how much they are willing to lend at each rate of interest. Here, the equilibrium rental (or interest) rate and the total level of capital demanded are determined by the interaction of (1) demand for financial capital by firms and (2) the supply of financial capital by investors (see Figure 3.1).

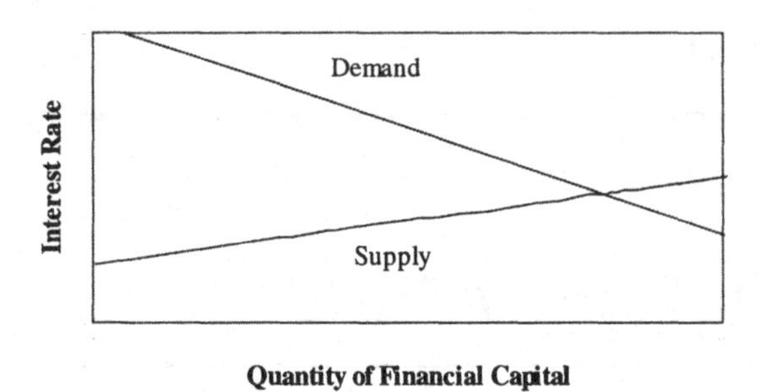

Figure 3.1. The market for financial capital.

From Chapter 2, we know that firms will demand more capital as its price falls and that investors will supply more capital as its price rises. A simple extension of the theory of profit-maximizing behavior states that firms will optimally demand a level of capital such that the increase in revenues that the capital produces is just equal to its price (this is its ***marginal revenue product***; see Pindyck and Rubinfeld, 2001, Chapters 14 and 15). Therefore, if the cost of capital rises (holding total productivity constant), the optimal level of capital used by the firm falls. What determines how much capital investors are willing to supply at each price?

As discussed in Chapter 2, we assume that individual investors attempt to maximize their well-being. Further, we will assume throughout this chapter that individuals prefer *certainty* to *uncertainty*. Given a choice between an uncertain outcome and an equivalent certain outcome, most individuals would prefer the equivalent certain outcome. For example, consider the choice between (1) receiving US$100 with probability one-half and $0 with probability one-half and (2) $50 with probability one. Most individuals would choose the latter.

We refer to ***uncertainty*** as ***risk*** and the preference for certainty as ***risk aversion***. This is discussed in Exercises 3.1–3.4. (Although one could make a distinction between risk and uncertainty, we will use them interchangeably in this chapter. More specifically, risk refers to situations with *known* probability distributions, whereas uncertainty refers to situations with *unknown* probability distributions. See Pindyck and Rubinfeld, 2001, p. 150.)

If investors were risk averse at each interest rate in the financial market, they would be willing to invest less in projects that are more risky. (We will broaden the definition of risk in Section 3.4.) Therefore, we should consider not one, but many, financial capital markets, in which projects with similar risks compete for investment funds. Because these markets are interrelated, we could plot the rental rate for capital in each market as a function of investment riskiness. In this view of the financial world, the role of the electric utility decision maker is to determine the risk class of each investment and the appropriate rate of return on capital for the appropriate risk class.

Until we develop an explicit relationship between risk and the interest rate, let us assume there is at least one *certain* investment: one investment that *always* pays a *known* return to the lender. This could be 90-day (e.g., U.S.) government securities.

We call the (nominal) rate of interest on the risk-free investment the ***risk-free interest rate***, R_f.

There are two components to the risk-free rate. One component compensates investors for *inflation* (i). The other component compensates them because they are unable to use the funds while they are being used by the borrower. This is the *real* risk-free interest rate, r_f. So the *nominal* risk-free interest rate (for example, the rate that the US government promises to pay on a 90-day security) is

$$1 + R_f = (1 + i) \cdot (1 + r_f) = 1 + i + r_f + (i \cdot r_f) \tag{3.1a}$$

$$R_f = i + r_f + (i \cdot r_f) \tag{3.1b}$$

$$R_f \cong i + r_f \tag{3.1c}$$

If inflation is low (e.g., less than 5%) and the real risk-free rate is low (e.g., less than 3%), then the last term in Equation (3.1a) will be small (e.g., $i \cdot r_f < 0.15\%$). So, we often simplify this relationship as $r_f = R_f - i$, i.e., the real risk-free rate is equal to the nominal risk-free rate minus the rate of inflation. However, one should *always* remember the last term ($i \cdot r_f$), particularly if the rate of inflation is high.

Finally, for all risky projects, investors will charge a ***risk premium*** (*RP*) to compensate them for the unknown rate of return:

1. The *real (risky) interest rate* is equal to the *real risk-free rate* plus a *real risk premium, RP*, $r = r_f + RP$.
2. The *nominal (risky) interest rate* is equal to the *nominal risk-free rate* plus a *nominal risk premium, RP**, which also compensates lenders for the riskiness of an unknown inflation rate, $R = R_f + RP^*$.

Throughout this chapter we will use *real* interest rates. In the next section we will assume that the *real* (risky) interest rate r is 10%, which is equivalent, for example, to *a risk-free rate*, $r_f = 5\%$ and a *risk premium*, $RP = 5\%$. If the rate of inflation is 10%, then the *nominal* (risky) interest rate, R, would be at least 20% with stable inflation: inflation (10%) + the risk premium (5%) + the real interest rate (5%) + ($i \cdot r_f = 0.25\%$). Given these definitions, we next discuss how firms choose between projects with similar risk. In Section 3.4, we discuss measuring risk.

3.2. NET PRESENT VALUE

In this section we present the most widely accepted method of project evaluation in modern finance: *net present value maximization*. We begin by defining and discussing ***discounting*** from the future into the present and from the present into the future. We then extend our discussion of profit-maximizing behavior from Chapter 2 to net present value maximizing behavior.

3.2.1. Discounting to the Present

The most important concept in finance is the "time value of money," i.e., that money in hand today has a different value than the same amount one year ago or one year from now. We will assume a perfect capital market, where the rental rate on finance capital ("money") is the same for the borrower or the lender, i.e., there are no transaction (e.g., banking) costs, so that the firm pays 10% per year to the investor. At this rate, $r = 10\%$, one dollar lent today (***present value,*** PV) will be worth \$1.10 in one year (***future value,*** FV):

$$FV = (1 + r) \cdot PV \tag{3.2a}$$

and

$$PV = [1/(1 + r)] \cdot FV \tag{3.2b}$$

We will refer to $[1/(1 + r)]$ or $(1 + r)^{-1}$ as the discount factor. Here, with $r = 10\%$, the discount factor is 0.9090, or about 0.91. We can extend this formula to two or more years:

$$\begin{aligned} PV_0 &= [1/(1 + r)] \cdot FV_1 \\ &= 1/[(1 + r) \cdot (1 + r)] \cdot FV_2 \end{aligned} \tag{3.3a}$$

$$PV_0 = [1/(1 + r)^t] \cdot FV_t, \quad t = 2 \tag{3.3b}$$

We follow the convention of designating the present period as period 0 ($t = 0$), the next future period as period 1 ($t = 1$), etc. We will assume that interest is paid on the *first day* of the next period. This can be extended to

$$FV_2 = (1 + r) \cdot FV_1 = (1 + r) \cdot (1 + r) \cdot PV_0 \tag{3.4a}$$

$$FV_t = (1 + r)^t \cdot PV_0 \tag{3.4b}$$

Here, we are using compounding: A return is earned on the return from a previous period. For example, if \$100 is lent today (period 0), in one year it is worth \$100 plus \$10, in two years it is

\$100 plus
\$ 10 (as a return in the first period) plus
\$ 10 (as a return in the second period) plus
\$ 1 (as a return on the return in the first period, equal to 10% of \$10), or
\$121.

If the project was riskier and financial institutions required a high risk premium, so that $r = 20\%$, in two years the lender would earn

$100 plus
$ 20 (as a return in the first period) plus
$ 20 (as a return in the second period) plus
$ 4 (as a return on the return in the first period, equal to 20% of $20), or
$144.

But if the project has no risk (see Exercise 3.3 on measuring risk), a risk-free rate of 5% would earn

$100 plus
$ 5 (as a return in the first period) plus
$ 5 (as a return in the second period) plus
$ 0.25 (as a return on the return in the first period, equal to 5% of $5), or
$110.25.

We can also discount by breaking the periods into subperiods, for example, into months. Under monthly discounting, the annual rate of 10% is divided by twelve, or 0.833% per month. With monthly compounding, the annual rate is slightly higher than with a 10% annual rate:

$$[1 + (10\%/12)]^{12} = 1.10471$$

or an effective interest rate of 10.47%. The appropriate monthly rate equivalent to an annual rate of 10% would be 0.797%, not 0.833%. Further, under daily discounting the annual rate would be divided by 365, or 0.0274%. Again, with compounding the annual rate would be higher than 10%:

$$[1 + (10\%/365)]^{365} = 1.10515$$

or an interest rate of 10.52%. At the limit we could continuously compound

$$[1 + (10\%/m)]^m = 1.10517 = e^{0.1} \text{ (or } \exp\{0.1\}) \text{ as } m \to \infty$$

Therefore, we can approximate subannual discounting with exponential constant (= 2.7183 . . .) to the power of the interest rate: e^r. Notice that monthly compounding approaches continuous compounding and daily compounding is almost indistinguishable from continuous compounding. When continuous compounding is appropriate, the mathematics of discounting can be greatly simplified.

Next, we consider a series of cash flows in each future period discounted to the present. For example, let A be a uniform amount (called an ***annuity***) invested each year at 10% for 3 years. What is the present value of this investment?

$$PV_0 = (1 + r)^{-1} \cdot A + (1 + r)^{-2} \cdot A + (1 + r)^{-3} \cdot A \qquad (3.5a)$$

This present value can be represented by the formula

$$PV_0 = A \cdot [(1+r)^T - 1]/[r \cdot (1+r)^T] \tag{3.5b}$$

where T is the last year. The inverse of the last part of Equation (3.5b), $[r\,(1+r)^T]/[(1+r)^T - 1]$, is the ***capital recovery factor***, *CRF*. Given an investment made in the present, what amount must be collected each year to recover the investment? It is

$$A = PV \cdot [r \cdot (1+r)^T]/[(1+r)^T - 1] \tag{3.6}$$

This is also known as the ***levelized capital cost***, because it is a level or uniform cost in each year. For example, if an investment of \$1000 is made today, what amount must be collected each year for 30 years to compensate the investor? Using Equation (3.6) with $r = 10\%$, \$106.08 must be collected each year for 30 years. Further, we can extend t into the future. As the number of periods increases, annuities approach a ***perpetuity***, i.e., to the equivalent of a payment made every period in the foreseeable future:

$$PV = A/r \qquad \text{and} \qquad A = PV \cdot r \tag{3.7}$$

For example, assuming r is 10% and the present value is \$1000, the annual payment would be \$100. So for long recovery periods, the simplified perpetuity approach yields values close to the exact formula. Although there are many other formulas to discount more complicated cash flows, such as increasing annual payments or changing rates of return, discounting future values into the present allows us to discuss net present value.

3.2.2. Net Present Value

In Chapter 2 we assumed that the firm maximized profit in each period. But firms, particularly those owning physical capital with long productive lives, such as electricity generation capacity, must make decisions based on future profits. We now extend our behavioral assumption from single-period profit maximization to multi-period profit maximization.

Assuming a competitive market for buying and selling firms, how much would a buyer offer the owner of the firm for the potential of receiving the firm's profits in each future period? Considering profits (*net cash flows*) in two future periods, $PR_t = (TR_t - TC_t)$, what is the present value of the firm (in period 0)?

$$\begin{aligned} NPV_0(PR_1, PR_2) &= PR_1 \cdot (1+r)^{-1} + PR_2 \cdot (1+r)^{-2} \\ &= (TR_1 - TC_1) \cdot (1+r)^{-1} + (TR_2 - TC_2) \cdot (1+r)^{-2} \end{aligned} \tag{3.8a}$$

where r is the buyer's real cost of capital. Or, more generally for T periods,

$$NPV_0(PR_1, \ldots, PR_T) = (TR_1 - TC_1) \cdot (1 + r)^{-1} + \ldots + (TR_T - TC_T) \cdot (1 + r)^{-T} \tag{3.8b}$$

This is the ***Net Present Value*** (NPV) of the firm, because it is *present value* of future revenues, *TR*, minus (net of) future costs, *TC*. Because we have assumed that these costs include a normal rate of return on invested capital, $NPV > 0$ implies positive (above normal) discounted profits.

This approach assumes that costs are incurred in each future period, but some costs, such as capital costs, must be spent in the current period so revenues can be received in the future. If the fixed cost is incurred in the first period and variable costs are spent in the future periods, then Equation (3.8b) becomes

$$NPV = -FC_0 + (TR_1 - VC_1) \cdot (1 + r)^{-1} + \ldots + (TR_T - VC_T) \cdot (1 + r)^{-T} \tag{3.9}$$

where TR_t and VC_t represent relevant positive and negative *cash flows* in each future period, discounted at the relevant cost of capital, *r*, to the present. If neither *real TR* nor *VC* change over time, then we can simplify Equation (3.9):

$$NPV = -FC_0 + (TR - VC) \cdot [(1 + r)^T - 1]/[r \cdot (1 + r)^T] \tag{3.10a}$$

$$= -FC_0 + (TR - VC) \cdot (1/CRF) \tag{3.10b}$$

We can extend the behavioral assumption of profit maximization to a multiperiod framework by assuming firms act to maximize their NPV. In particular, they invest in all projects (e.g., power plants) with positive NPV, because all projects with positive NPV imply above-normal discounted profits. This is the *Net Present Value Rule*. However, given constraints on their ability to manage multiple projects, we assume that (1) firms rank possible projects by NPV and (2) begin by investing in the project with the highest NPV. Following this strategy, firm managers maximize NPV for the firm's owners. (On electric utility reaction to projects with negative NPV, see Rothwell, 1997. On the distribution of uncertain NPVs, see Rothwell, 2001.)

3.2.3. Assessing Cash Flows under the Net Present Value Rule

Before comparing the NPV method with other project evaluation methods, we should discuss a few issues related to appropriately assessing *cash flows* in an NPV analysis (see Brealey and Myers, 2000, Chapter 6).

The first is the definition of *working capital* (to be discussed in Chapter 4). Working capital is the difference between short-term ***assets*** (benefits to the firm) and short-term ***liabilities*** (costs to the firm):

1. *Short-Term Assets* include customer's unpaid bills (accounts receivable), cash, and the value of raw materials and finished goods inventories.
2. *Short-Term Liabilities* include the firm's unpaid bills (accounts payable).

Regarding NPV analysis, changes in working capital should be included as a cash flow.

Second is the definition of ***depreciation***. Depreciation is used in tax and regulatory accounts. For tax purposes, depreciation reduces the (book) value of an asset and can be counted against income, reducing tax liabilities. (The book value is the value of an asset in the firm's accounting "books.") For regulatory purposes it provides the regulated firm with income from customers to pay for capital investments that must be replaced at the end of their useful life. The physical counterpart to depreciation is wear-and-tear. Depreciation can also include economic or technical obsolescence. However, when incorporating depreciation or other tax effects, discounting after-tax cash flows requires the use of after-tax discount rates.

For tax purposes there are several methods of calculating depreciation. The *straight-line depreciation rate* is equal to the inverse of the life of the asset. For example, if a power plant has a 25-year life, then the straight-line depreciation rate is 1/25 or 4% per year. Many regulators use the straight-line method.

However, the tax authority (to encourage investment) can allow firms to accelerate their depreciation. One method of accelerated depreciation is double-declining balance. This method doubles the simple depreciation amount applied to the remaining book value of the asset. For example, if the simple depreciation rate is 4%, the double-declining balance rate would be 8% on the remaining book value (after depreciation). Double-declining balance and other accelerated depreciation methods increase depreciation in the early life of the asset and decrease depreciation in the later years. Therefore, *after-tax* income is higher in the early years than with straight-line depreciation and is lower in the later years. NPV maximizers prefer accelerated depreciation because the present value of income is greater the closer it is to the present. Unfortunately, if the tax authority is using one form of depreciation and the regulator is using another form, two sets of accounts must be kept. (The stock market regulator might require a third set of accounts for the shareholders. Also, there might be a fourth set for the firm's management to make investment and production decisions.)

To complete our discussion of depreciation, we introduce the concept of *salvage value*. This is the value of an asset at the end of its useful life. Generally, the salvage value is assumed to be positive. But the salvage value could be negative. For example, if a nuclear power plant must be decommissioned (decontaminated and dismantled), and the cost of decommissioning is $300M, then the salvage value would be –$300M. The cost of decommissioning could be financed through depreciation. However, tax and regulatory authorities usually treat decommissioning accounting separately from depreciation accounting (see Pasqualetti and Rothwell, 1991).

Because we have acknowledged the tax collector, we conclude this section by discussing the influence of taxes on the NPV calculation. Because after-tax cash flows are income to an NPV maximizer, taxes should be subtracted from profits before they are discounted to the present. For example, if the tax rate on income (revenues minus costs) is 50%, then annual revenues to the owner of an electric generator are reduced by half. If customers contribute an amount equal to the depreciation allowed by the tax authorities and another amount equal to the income taxes, the NPV analysis is unchanged under regulation. If electricity generation is deregulated (see Chapter 5) and firms must subtract taxes from after-depreciation income, the NPV calculation will change. Therefore, the firm must be fully aware of the tax collector and calculate the NPV on projects accordingly.

3.3. ALTERNATIVE METHODS OF PROJECT EVALUATION

Although net present value analysis is the most widely accepted form of project evaluation in modern finance, there are at least three others that are used and can be compared with the NPV approach:

1. Payback analysis
2. Average return on book value
3. Internal rate of return

We give an introduction to each of these here, but because each of these methods has problems that NPV does not have, we will not discuss them completely. Interested readers should consult Brealey and Myers (2000, Chapter 5).

3.3.1. Payback Analysis

The *Payback Rule* is commonly used in small businesses as a rule-of-thumb for making routine investment decisions, such as whether to install fluorescent light bulbs to reduce the cost of electricity. The rule is simple to use: make those investments that pay back the cost of the initial investment after some arbitrarily chosen (payback) period, e.g., 2 years. For example, if

1. The cost of a fluorescent bulb is $10 *greater than* the cost of the nonfluorescent bulb
2. The *reduction* in electricity is 100 watts
3. The bulb is used 10 hours per day
4. The price of electricity is $0.10/kWh (or $100/MWh)

then the payback period is 100 days: $10 (the cost of the bulb) = $0.10/kWh · 100 watts · 10 hours · 100 days. If the payback period is 2 years, then the business owner buys the bulb. Notice, however, that whereas this method is easy to use, the business owner is not considering the savings after 100 days. Many fluorescent bulbs last for years, so the savings over the life of the bulb should be considered. If the payback period is too short, then some investments with a positive NPV will be ignored. If the payback period is too long, some investments with a negative NPV will be accepted. If the net revenues over the life of the project are uniform (as in this example), the NPV-equivalent payback period is equal to

$$(1/r) - 1/[r \cdot (1 + r)^T] \tag{3.11}$$

where T is the life of the project. If $r = 10\%$ and $T = 3$ years, then the NPV-equivalent payback period is about 2.5 years. In this example, the business owner would come to the same conclusion using the payback rule or the NPV maximization rule, but with complicated investment decisions, the Payback Rule is too simplistic.

3.3.2. Average Return on Book Value

Another project evaluation approach considers the ratio of (1) the average income over the life of the investment to (2) the average book value of the investment (after depreciation). If the ratio is above the average accounting rate of return for the firm, then the investment is made. For example, if the business owner is contemplating buying a computer system with a 3-year economic life (after the end of 3 years the computer system is essentially worthless) for $3000, then the average book value (assuming straight-line depreciation) is

- $3000 at the beginning of the first year
- $2000 at the beginning of the second year
- $1000 at the beginning of the third year
- Then $2000 is the average for 3 years

Assuming that the computer saves the business owner $1000 per year in labor costs (after training), the average return on book value is 50%, which is above the firm's accounting rate of return, so the business owner buys the computer.

But this approach ignores the fact that the computer costs $3000 in the present and any discounting reduces the value of future savings. If the cost of capital is 10%, then the value of the labor savings in the second year discounted to the present is $909 and its value in the third year is $826. The NPV at the time of the purchase is $1000 (first year) + $909 + $826 – $3000, or –$265. This is not a good investment under NPV maximization with $r = 10\%$. (Also, if the payback period were 2 years, the business owner would not invest, but would invest if the payback period were 3 years.) The problem with the average rate of return (and similar methods) is that it does not consider actual cash flows.

3.3.3. Internal Rate of Return

The third approach is the internal rate of return method. Under it, the decision maker calculates the rate of return that yields NPV = 0 and compares this rate with the firm's (internal) rate of return. If this *Internal Rate of Return* (*IRR*) is greater than the cost of capital, then the firm selects the project.

For example, let's assume that Equation (3.10a) and a 10% cost of capital are appropriate, then the IRR method is (1) solve the following equation for *IRR*:

tribution, the probability of finding an observation between one standard deviation above the mean and one standard deviation below the mean is about two-thirds (68%). The probability of finding an observation between two standard deviations above and below the mean is about 95%. The probability between three standard deviations above and below the mean is about 99.75%. (Total probability is 100%.)

In our example, if the four observations on the firm's rate of return were drawn from a normal distribution then the probability of observing a rate of return between 1.5% (= 5% – 3.5%) and 8.5% (= 5% + 3.5%) is 68%. The probability of observing a rate of return between –5.5% (= 5% – 3 · 3.5%) and 15.5% (= 5% + 3 · 3.5%) is almost one (99.75%). This does not imply that we could not find observations outside these bounds. With only four observations, our estimate of the mean and standard deviation is not robust. Generally, we need more than 30 observations to accurately estimate the characteristics of a normal distribution.

3.4.1. Financial Instruments

Although an investor can invest in many types of financial instruments, we will limit our discussion to three types of instruments: *government securities, corporate bonds,* and *corporate stocks*. Government securities or government bonds are issued by all forms of governments to finance their expenditures. For example, a federal government can sell long-term bonds to make up the difference between current expenditures and tax revenues, or a local government can sell bonds to build schools or other infrastructure. Although there can be risks associated with the returns on government bonds, generally, stable governments pay investors for holding short-term securities, either out of tax revenues or out of revenues from the sale of longer-term bonds.

For example, 90-day U.S. government securities (Treasury Bills) are considered risk-free. Therefore, it is possible to determine the current *nominal risk-free interest rate*, R_f, by looking at the rates of return offered to investors published in the financial sections of newspapers. Some governments issue inflation-adjusted securities, such as the US government's "Treasury Inflation Bonds." The rate of return to investors is free of inflation or inflation risk. The interest rate on these instruments is (theoretically) equal to the *real risk-free interest rate*, r_f. Therefore, the *expected* inflation rate equals the difference between these two interest rates, as discussed in Section 3.1.

Financial instruments similar to government bonds are bonds issued by non-government entities, such as corporate firms or joint-stock companies. These firms usually issue at least two types of financial securities:

- ***bonds*** are a form of *debt* and pay a fixed payment (the coupon) in each period
- ***stocks*** (sometimes known as ordinary *shares*) are a form of *equity* and pay ***dividends***

Although there can be many forms of bonds and stocks with different risk characteristics, the difference between bonds and stocks is the risk associated with claims on the firm's earnings. When there are earnings, bondholders are paid before stockholders. When there are no earnings, bondholders can claim the property of the firm. This can take the form of forcing the firm to declare bankruptcy and allowing a court to sell the firm's property to pay the bondholders. This is a right of a creditor. For example, another form of debt is mortgage debt backed directly with property, such as a house. If the mortgagor does not pay the mortgagee, the mortgagee can seize the property, sell it, pay the mortgage, and return the difference, if any, to the mortgagor.

Equities, on the other hand, are a form of joint ownership. The equity investor shares in the returns of the company with no guarantee of a return. In some years profits are high and in other years they are low or zero. The equity holder does not have the right to seize property, but has the right to participate in the management of the company (such as voting for the Board of Directors) and to a share of the proceeds if the firm's property is sold and creditors have been paid. In a limited liability company, the loss of the equity holders is limited to their investment. If the investor does not have limited liability, the investor's property can also be seized to pay the debts of the bankrupt firm.

The returns on corporate bonds have a wider variance than returns on bonds issued by governments. Although the bond pays a fixed amount in each period, and thus seems certain, there is no guarantee that the issuer of the bond will pay the fixed amount. The firm might default on its bond payments. (Governments can also default on their bonds.) Although bond payments are fixed, the prices of the bonds in financial markets can change. With a fixed payment, if the price of the bond *falls*, the implied interest rate *rises*.

For example, assume the price of a bond is $1000 and pays $100 per year in perpetuity. The rate of interest on the bond would be $100/$1000 or $r = 10\%$. If the price of the bond drops to $900 (and continues to pay $100 per year as stated on the bond), then the rate of interest would be $100/$900 or $r \cong 11\%$. Therefore, given the fixed payment, as the price of bonds fall, the interest rate on bonds rises. Or if interest rates on government securities rise, the price of a corporate bond falls, given a fixed payment. Because regulated electric utilities rarely default on their bond payments (although there is a possibility, see Chapter 6 on the bankruptcy of Pacific Gas & Electric in California), interest rates on bonds expected by investors issued by electric utilities are usually among the lowest rates of all corporate bonds.

Because corporate equities ("stocks" or "shares") are not backed by the property of the firm or the tax authority of a government, they are far riskier than government or corporate bonds. In one year, the firm might pay a dividend of 5% on investments, pay 10% the next year, nothing in the following year, and 5% in the fourth year, and so on. Because of the uncertainty of the return, shareholders expect a higher rate of return than bondholders do. Often, these shares are traded in financial markets, such as the New York Stock Exchange.

The rate of return depends on the share price and the expected dividend. If investors expect a lower dividend, the price of the share decreases, thus raising the rate of return on the share. For example, if an electric utility announces that it will pay $1 per share in a particular quarter (of a year), and the share is trading at $40, then the rate of return on the share is 2.4% per quarter or about 10% for the year. If investors were expecting a higher return, they might sell their shares, increasing the supply of the firm's shares in the market. So the price of the share falls. If it falls to $35, then the $1 dividend yields a quarterly rate of return of 2.8% or about 11.6% per year.

Of course, this is a simplified explanation of how investors behave in stock markets, because they also have expectations about future dividends and future changes in share prices. Generally, firms in the same risk category can expect to pay the same rate on equity capital in financial markets. Therefore, when trying to determine a firm's cost of capital, one needs to determine the risk class of the firm.

Although it is difficult for individual investors to determine the riskiness of corporate debt and equity, there are rating services that grade these financial instruments (Brealey and Myers, 2000, Chapter 23). For example, Moody's Investor Services grades corporate bonds. These grades range from Aaa for bonds with the lowest risk (and lowest expected return) to C for bonds with the highest risk (and the highest expected return). Other investor services grade bonds issued by countries or firms in a country. Once an electric utility's bonds have been graded, i.e., assigned to a risk category, it is easy to determine the utilities interest rate on debt. Determining expectations regarding the cost of equity is more difficult. It involves (1) determining the risk category of the utility's equities and (2) determining what financial markets are charging firms in that risk category. These issues are discussed in the exercises below.

To summarize this discussion, Figure 3.3 shows the historic relationship between the cost of capital and risk. The vertical axis shows the historic annual average rate of return and the horizontal axis shows risk, measured by the annual standard deviation of the rates (Brealey and Myers, 2000, Chapter 7). As the riskiness of a class of financial instruments increases, so does the return expected by investors.

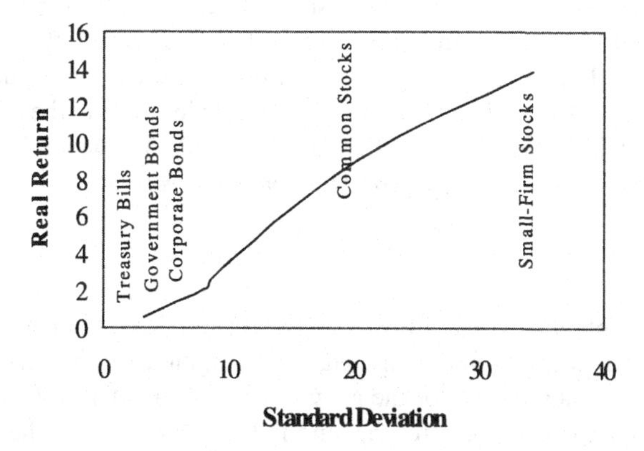

Figure 3.3. Historic risks and returns. (After Brealey and Myers, 2000, pp. 156 and 164.)

3.4.2. Capital Structure and the Cost of Capital

Because the firm can use both debt and equity financing, its cost of capital is a weighted average of the rate of interest on debt and the expected rate of return on equity. The nominal ***Weighted Average Cost of Capital***, *WACC*, is

$$WACC = [Debt/(Debt + Equity)] \cdot R_{debt} + [Equity/(Debt + Equity)] \cdot R_{equity} \tag{3.15}$$

Assume 70% of an electric utility's financing is in corporate bonds and 30% is in corporate equities. If $R_{debt} = 7\%$ and $R_{equity} = 12\%$, then the *WACC* is 8.5%.

However, using the *WACC* to discount cash flows for purposes of NPV calculations is only appropriate if the costs of debt financing have been ignored in the calculation of cash flows (i.e., not subtracted from cash flows). If the costs of debt financing have been included in the calculation of cash flows (i.e., subtracted from the cash flow), then the appropriate discount rate is the equity rate.

Although one might conclude from Equation (3.15) that increasing the percentage of debt in the firm's total capitalization would lower the *WACC*, debt financing also increases the probability of default. This is because creditors *must* be paid; if there is a decrease in earnings, the firm might not pay all creditors. Therefore, bondholders will demand a higher rate of return as the percentage of debt rises (and riskiness increases). Increasing debt is *financial leverage*, because equity holders leverage their small contribution with large contributions from debt holders (Brealey and Myers, 2000, Chapter 19).

EXERCISE 3.1. RISK AND DIVERSIFICATION

To minimize risk and maximize expected return, individuals invest in a diverse set of financial instruments. This exercise explores how risk can be reduced by investing in a *portfolio* of assets. We focus on a portfolio of common corporate stocks selected by Dow Jones & Company, publishers of the *Wall Street Journal*. Dow Jones tracks three portfolios of common stocks: industrials, transportation, and utilities. The Dow Jones Industrial Average is a portfolio of 30 common stocks of large industrial corporations headquartered in the US. Table 3.1 lists the *monthly* percentage returns for 1997 for 12 companies listed in Table 3.2, which lists the company name and the common stock symbol. Here, *nominal return*,

$$R_t = (P_{t+1} - P_t)/P_t \tag{3.16}$$

where P_t is the period-t price of the stock (with capitalized dividends). Also included in Table 3.1 is the (value-weighted, i.e., weighted by the total value of each firm's equity) average return for the New York Stock Exchange (NYSE).

Investors compare average returns and risk for each stock. The average, or *expected value* of the return, $E(R_t)$, is defined in Equation (3.17):

Table 3.1. 1997 Monthly Percentage Returns for 12 Stocks in the Dow Jones Industrial Average

1997	NYSE	AXP	BA	CHV	C	KO	DIS	XON	GE	HWP	IBM	MCD	WMT
Jan	5.3	10.0	0.6	2.1	5.7	10.0	4.6	5.7	4.7	4.7	3.5	0.3	4.4
Feb	−0.1	5.6	−4.8	−2.0	−2.5	5.4	1.9	−2.5	−0.6	6.7	−8.1	−4.8	11.1
March	−4.4	−8.8	−3.1	7.9	−10.6	−8.4	−1.9	7.5	−3.0	−4.7	−4.5	9.2	5.9
April	4.3	10.2	0.0	−1.6	0.0	14.1	12.4	5.1	11.8	−1.6	16.9	13.2	0.9
May	7.1	5.7	7.1	3.0	6.3	7.7	0.2	5.4	8.8	−1.9	7.9	−5.9	6.2
June	4.4	7.2	0.7	5.6	4.4	−0.5	−2.0	3.4	7.7	9.0	4.3	−3.9	13.4
July	7.6	12.7	10.6	6.8	13.1	1.7	0.9	4.9	8.3	25.0	17.2	11.3	10.9
Aug	−3.7	−7.2	−6.9	−1.2	−5.5	−17.1	−5.0	−4.1	−10.8	−12.1	−3.9	−11.8	−5.3
Sept	5.8	5.3	−0.1	7.3	5.9	6.7	5.0	4.7	9.2	13.3	4.6	0.7	3.4
Oct	−3.4	−4.5	−11.8	−0.2	−4.2	−7.2	2.3	−4.1	−5.1	−11.4	−7.1	−5.9	−4.4
Nov	3.1	1.1	11.0	−2.6	−2.7	10.6	15.3	0.0	14.3	−0.8	11.4	8.4	14.5
Dec	1.8	13.4	−7.9	−4.0	3.7	6.7	4.3	0.3	−0.3	2.3	−4.5	−1.5	−1.4

Note: Data does not necessarily refute actual returns.

$$\mathrm{E}(R_t) = (1/\mathrm{T}) \cdot \Sigma\, R_t \qquad \text{for } t = 1, \ldots, T \qquad (3.17)$$

The most common measure of variation is the *standard deviation*, $\mathrm{SD}(R_t)$, which is the square root of the *variance*, $\mathrm{VAR}(R_t)$:

$$\mathrm{VAR}(R_t) = [1/(T-1)] \cdot \Sigma\, [R_t - \mathrm{E}(R_t)]^2 \qquad \text{for } t = 1, \ldots, T \qquad (3.18)$$

Table 3.2 lists percentage values of $\mathrm{E}(R_t)$, $\mathrm{SD}(R_t)$, and $\mathrm{VAR}(R_t)$ for the 30 stocks in 1997. (As mentioned above, an unbiased estimator for sample variance accounts for the degree of freedom lost in calculating the sample mean using $1/[T-1]$ in place of $1/T$.)

Notice in Table 3.2 that the standard deviation of the returns to the market (NYSE) is lower than all but one of the standard deviations of the individual stocks. This is because the variance of a portfolio depends on (1) the variances of the stocks in the portfolio and (2) the covariances between these stocks. ***Covariance*** between the returns on two stocks, j and k, is

$$\mathrm{COV}(R_{jt}, R_{kt}) = [1/(T-1)] \cdot \Sigma\, [R_{jt} - \mathrm{E}(R_{jt})]\, [R_{kt} - \mathrm{E}(R_{kt})] \qquad \text{for } t = 1, \ldots, T \qquad (3.19\text{a})$$

Although portfolio variance is a function of covariance between the stocks in the portfolio, it is easier to work with ***correlation***, defined as

$$\mathrm{CORR}(R_{jt}, R_{kt}) = \mathrm{COV}(R_{jt}, R_{kt})/[\mathrm{SD}(R_{jt}) \cdot \mathrm{SD}(R_{kt})] \qquad (3.19\text{b})$$

Positive correlation implies that the two stocks move up and down together. Negative correlation implies that the two stocks move in opposite directions, i.e., when one goes up, the other goes down. If the correlation coefficient is 1, the two stocks

Table 3.2. The Issuers of the Stocks in the Dow Jones Industrial Average

Symbol	Company	E(*R*)	SD(*R*)	VAR(*R*)
NYSE	New York Stock Exchange	2.32	4.28	0.18
AA	Alcoa	1.22	7.84	0.61
ALD	Allied Signal	1.62	7.48	0.56
AXP	American Express	4.24	7.53	0.57
BA	Boeing	–0.38	7.16	0.51
C	Citigroup	1.13	6.54	0.43
CAT	Caterpillar	2.47	6.19	0.38
CHV	Chevron	1.77	4.28	0.18
DD	Du Pont	2.49	7.72	0.60
DIS	Disney	3.16	5.81	0.34
EK	Eastman Kodak	–1.84	7.67	0.59
GE	General Electric	3.75	7.62	0.58
GM	General Motors	1.62	5.13	0.26
GT	Goodyear Tire	2.15	6.46	0.42
HWP	Hewlett Packard	2.37	10.42	1.09
IBM	IBM	3.14	8.94	0.80
IP	International Paper	1.14	9.72	0.95
JNJ	Johnson & Johnson	2.79	8.19	0.67
JPM	J P Morgan	1.63	5.41	0.29
KO	Coca Cola	2.47	9.21	0.85
MCD	McDonalds	0.77	7.98	0.64
MMM	3M (Minn. Mining & Mfg.)	0.40	7.97	0.64
MO	Philip Morris	2.16	7.99	0.64
MRK	Merck	2.96	9.25	0.85
PG	Procter & Gamble	3.66	6.96	0.48
S	Sears Roebuck	0.71	11.94	1.42
T	AT&T	3.94	8.44	0.71
UK	Union Carbide	0.87	8.58	0.74
UTX	United Technologies	1.12	6.45	0.42
WMT	WalMart Stores	4.96	6.68	0.45
XON	Exxon	2.18	4.09	0.17

Note: All values are in percentages. Data does not necessarily refute actual returns.

are perfectly positively correlated. If the correlation coefficient is 0, then the two stocks are independent. If the correlation coefficient is –1, the two stocks are perfectly negatively correlated.

For example, the correlation between American Express and Citigroup is +0.8 (or 80%). The stocks of these two financial services companies move together most of the time. But between American Express and Chevron the correlation coefficient is –0.1 (or –10%, American Express and Chevron, an oil company, although negatively related, are almost independent).

We now consider portfolios of assets. The *expected return of a portfolio* of two stocks is

$$\text{E(portfolio return)} = x \cdot \text{E}(R_{jt}) + (1 - x) \cdot \text{E}(R_{kt}) \tag{3.20}$$

where x is the proportion of value of the portfolio invested in one stock and $(1 - x)$ is the proportion of value of the portfolio invested in the other stock. The *variance of a portfolio* of two stocks is

$$\begin{aligned} \text{VAR(portfolio return)} &= x^2 \cdot \text{VAR}(R_{jt}) + (1 - x)^2 \cdot \text{VAR}(R_{kt}) \\ &\quad + 2 \cdot x \cdot (1 - x) \cdot \text{COV}(R_{jt,}\ R_{kt}), \\ &= x^2 \cdot \text{VAR}(R_{jt}) + (1 - x)^2 \cdot \text{VAR}(R_{kt}) \\ &\quad + 2 \cdot x \cdot (1 - x) \cdot \text{CORR}(R_{jt,}\ R_{kt}) \cdot \text{SD}(R_{jt}) \cdot \text{SD}(R_{kt}). \end{aligned} \tag{3.21}$$

For example, in an equally weighted portfolio of American Express and Chevron,

1. The expected return would be 3% =

$$(0.5) \cdot (4.24\%) + (0.5) \cdot (1.77\%)$$

2. The variance would be 0.179% =

$$(0.5)^2 \cdot (7.53\%)^2 + (0.5)^2 \cdot (4.28\%)^2 + 2 \cdot 0.5 \cdot 0.5 \cdot -0.1 \cdot 7.53\% \cdot 4.28\%$$

3. The standard deviation would be 4.14%, which is less than the standard deviations of either of the two stocks because of the negative correlation between the two returns.

By varying the proportions of stocks in the portfolio, the investor can find optimal combinations that minimize risk for each level of expected return. This is the *portfolio frontier*. Further, the investor can divide assets between the risk-free asset (e.g., a short-term government bond) and an optimal portfolio. If the investor can lend or borrow funds at the risk-free rate to purchase shares of the risk-free asset and an optimal ***market portfolio*** (a portfolio of equities, such as a portfolio of the Dow Jones Industrials), the resulting combinations will yield the highest return for *each* level of risk.

For example, assume the point [E(R) = 2.9%, SD(R) = 4.0%] is on the portfolio frontier. A weighted combination of this investment opportunity and the risk-free asset can be represented by a line. Assuming the line has an intercept equal to a risk-free rate of 0.5% and it passes through (2.9%, 4.0%) at 100% of the risky asset, it can be represented as

$$\text{E}(R) = 0.5 + 0.6025 \cdot \text{SD}(R)$$

This line can be plotted as a tangency on the portfolio frontier as in Figure 3.4, or alone, as in Figure 3.5. This line represents the highest rate of return for *each* level of risk, given the parameters of the problem.

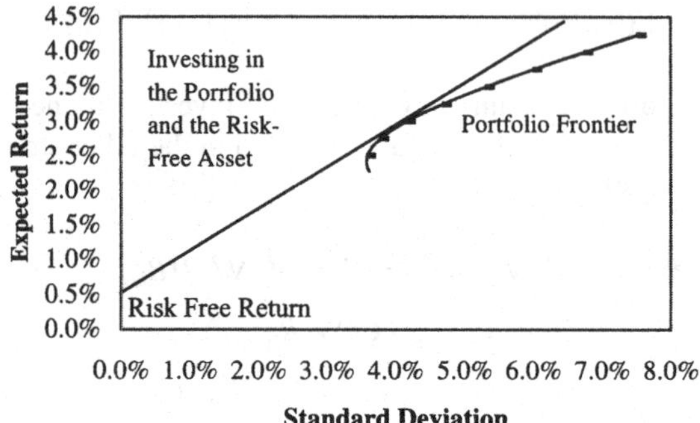

Figure 3.4. Efficient portfolios with borrowing and lending at the risk-free rate.

3.1.1. Plot $E(R_{jt})$ against $SD(R_{jt})$ with data from Table 3.2.

3.1.2. Calculate the expected return and standard deviation for a portfolio of American Express and Chevron at x = 100%, 90%, 80%, . . . , 20%, 10%, and 0% from the data in Table 3.1. Plot the portfolio's expected return as a function of the portfolio's standard deviation. (In later exercises, assume that this portfolio represents an optimal portfolio for the Dow Jones in 1997.)

EXERCISE 3.2. RISK AVERSION

To describe how investors evaluate risk, economics proposes a theory of risk aversion. It is based on the economic description of how individuals make choices between uncertain alternatives. The microeconomic theory of demand assumes that consumers (or investors) purchase goods and services (or financial instruments) to

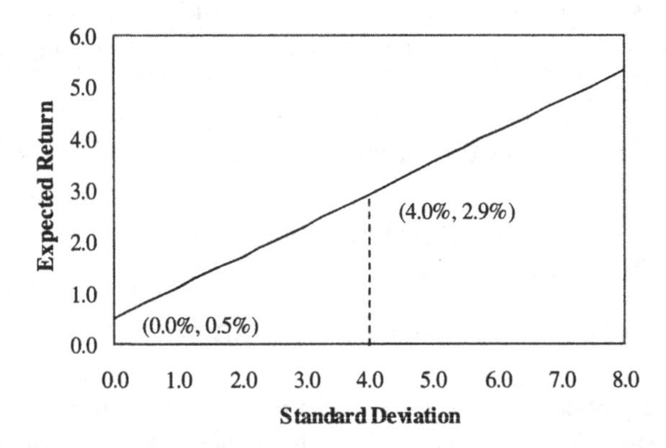

Figure 3.5. A portfolio of the risk-free asset and the market portfolio.

maximize their well-being or "utility" (welfare). (Here, "utility" is a term used by economists. It was inspired by the philosophical and political movement known as utilitarianism.)

Because it is impossible to compare well-being among individuals, *utility theory* is more abstract than *production theory*, but it has a similar framework. Like firms that maximize output subject to their production possibilities, in utility theory individuals attempt to maximize their utility, U (measured by an ordinal index), subject to their income. Solutions to this constrained maximization problem give rise to individual demand functions (quantities demanded by the individual at each price), which can be summed up to market demand functions.

A reduced form of this maximization problem focuses on the individual's utility of wealth, W. How do individuals rank different levels of wealth? We refer to $U(W)$ as the ***utility function*** of wealth. What is a reasonable form for this function? $U(W_j)$ represents the value that the individual places on W_j. We want a utility function such that when $W_1 > W_2$, then $U(W_1) > U(W_2)$. But if W_1 and W_2 are uncertain, as they would be if wealth involved financial instruments, then how would individuals compare uncertain levels of wealth? For example, how would an individual compare the following choices?

1. A return of $5,000 half the time or a return of $20,000 half the time
2. A return of $12,500 all of the time

To help make this comparison, we specify the utility function as a (natural) logarithmic function of wealth (we will consider other functional forms below):

$$U(W) = \ln W \tag{3.22}$$

Figure 3.6 represents utility as a function of wealth in $1000s.

If W is equal to 5 half the time and 20 half the time (measured in thousands of dollars), then the ***expected utility***, E(U), would be

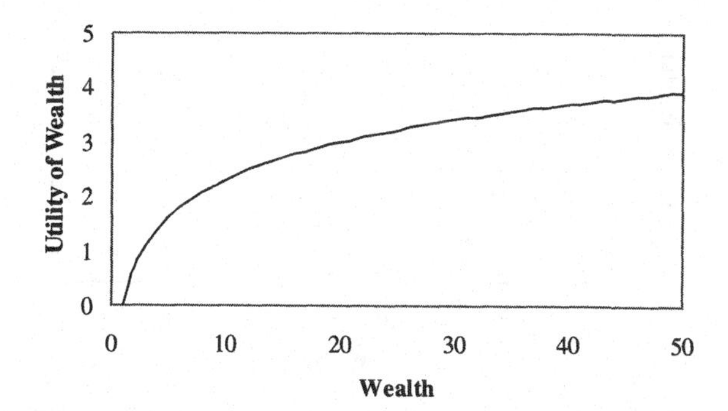

Figure 3.6. Utility as a logarithmic function of wealth.

$$\begin{aligned} E[U(W)] &= 0.5 \cdot U(5) + 0.5 \cdot U(20) \\ &= 0.5 \cdot \ln(5) + 0.5 \cdot \ln(20) \\ &= 0.5 \cdot 1.61 + 0.5 \cdot 3.00 \quad = 2.3 \end{aligned} \tag{3.23}$$

On the other hand, the *utility of the expected wealth* [E(W) = 12.5] would be

$$\begin{aligned} U[E(W)] &= U[0.5 \cdot 5 + 0.5 \cdot 20] \\ &= U[12.5] \qquad = 2.5 \end{aligned} \tag{3.24}$$

With logarithmic utility, the *utility of the expected value* of the certain choice is greater than the *expected utility* of the uncertain choice: $U[E(W)] > E[U(W)]$. Here, the individual would prefer a *certain outcome* (= 2.5) to an *uncertain outcome* (= 2.3), even though the *certain outcome* is equal to the expected value of the *uncertain outcome*. To describe this behavior, we define the following:

- If $U[E(W)] > E[U(W)]$, then the individual is *risk averse*
- If $U[E(W)] = E[U(W)]$, then the individual is *risk neutral*
- If $U[E(W)] < E[U(W)]$, then the individual is *risk preferring*

Risk aversion, where $U[E(W)] > E[U(W)]$, is portrayed in Figure 3.7.

The *risk premium*, *RP* (Figure 3.8), equates expected utility and the utility of the expected value:

$$E[U(W)] = U[E(W) - RP] \tag{3.25}$$

The risk premium is the value that makes the individual indifferent when choosing between a certain outcome and an uncertain outcome. In the above example,

$$RP = E(W) - \exp\{E[U(W)]\} = 12.5 - 10 = 2.5 \Rightarrow \$2500$$

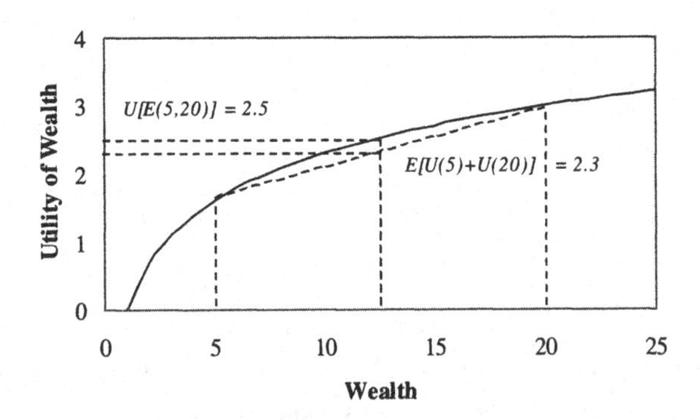

Figure 3.7. Utility of expected wealth versus expected utility.

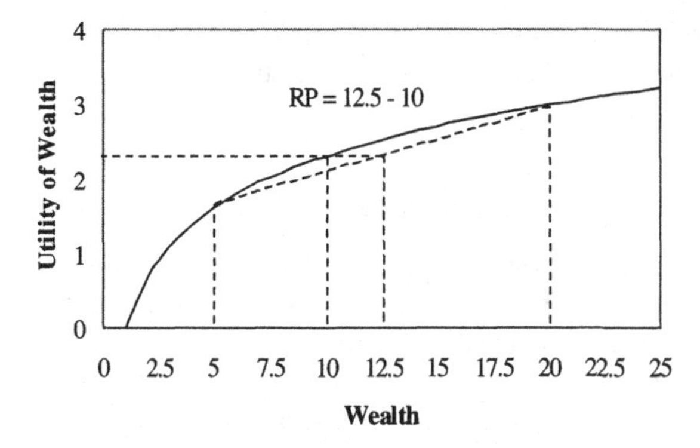

Figure 3.8. Calculating the risk premium.

Also, we would like to know how an individual's risk aversion changes with the level of the individual's wealth. A wealthy individual might not be as risk averse to a $10,000 gamble as an individual with an average income. To measure risk aversion as a function of wealth, we define *Absolute Risk Aversion*, *ARA*, as

$$ARA = -(\mathrm{d}^2U/\mathrm{d}W^2)/(\mathrm{d}U/\mathrm{d}W) \tag{3.26}$$

where $(\mathrm{d}U/\mathrm{d}W)$ is the first derivative of utility with respect to wealth and $(\mathrm{d}^2U/\mathrm{d}W^2)$ is the second derivative. For example, with logarithmic utility

$$\mathrm{d}U/\mathrm{d}W = 1/W$$

and

$$\mathrm{d}^2U/\mathrm{d}W^2 = -1/W^2$$

so

$$ARA = 1/W$$

With the logarithmic utility function, absolute risk aversion is an inverse function of wealth; as wealth increases, the absolute level of risk aversion declines: $\mathrm{d}ARA/\mathrm{d}W < 0$.

On the other hand, someone with a net worth of $10,000 might feel the same aversion to a $100 gamble as someone with a net worth of $10,000,000 evaluating a $100,000 gamble. To compensate for levels of wealth, *Relative Risk Aversion*, *RRA*, is defined as

$$RRA = -W(\mathrm{d}^2U/\mathrm{d}W^2)/(\mathrm{d}U/\mathrm{d}W) = -W \cdot ARA \tag{3.27}$$

For example, with logarithmic utility, $RRA = W \cdot ARA = W \cdot (1/W) = 1$, relative risk aversion is constant and $\mathrm{d}RRA/\mathrm{d}W = 0$, i.e., relative risk aversion does not change with wealth, whereas absolute risk aversion does.

3.2.1. The Quadratic Utility Function

With these definitions of risk aversion, absolute risk aversion, and relative risk aversion, the remainder of this exercise explores the quadratic form of the utility function. We will define reasonable criteria by which to judge the appropriateness of a specific functional form for the utility function. Then we explore how well the quadratic utility function approximates risk aversion. The quadratic utility function is

$$U(W) = a + b \cdot W + c \cdot W^2 \tag{3.28}$$

The first criterion for a well-behaved utility function is whether utility is increasing in wealth. But a quadratic function has a single maximum or minimum. For a single maximum (so that utility increases with increases in wealth), the second derivative should be less than zero. Determine the first and second derivatives of this quadratic utility function. What must be the sign of c to ensure that utility increases with wealth? Given this solution, what conditions must be satisfied to ensure that $\mathrm{d}U(W)/\mathrm{d}W > 0$? Given these requirements on the signs of b and c, calculate the absolute and relative risk aversion for the quadratic utility function. What additional restrictions should be placed on the parameters of the quadratic utility function to ensure reasonable behavior in absolute and relative risk aversion?

3.2.2. The Mean-Variance Approximation

Because the quadratic utility function has limitations, what are its appropriate applications? Assume that the utility function, $U(W)$, is not necessarily quadratic. A Taylor series expansion of $U(W)$ around some point x would yield

$$\begin{aligned} U(W) = {} & U(x) \cdot (W - x)^0 \\ & + [U'(x)/1] \cdot (W - x)^1 \\ & + [U''(x)/1 \cdot 2] \cdot (W - x)^2 \\ & + [U'''(x)/1 \cdot 2 \cdot 3] \cdot (W - x)^3 + \ldots \end{aligned} \tag{3.29a}$$

where U' is the first derivative of the utility function with respect to x, U'' is the second derivative, etc. The Taylor series expansion can approximate any function to high degrees of accuracy by extending the remainder. Because the derivatives of the utility function are constants, let $U(x) = a$, $U'(x) = b$, $U''(x)/2 = c$, and $U'''(x)/6 = d$. So,

$$U(W) = a \cdot (W - x)^0 + b \cdot (W - x)^1 + c \cdot (W - x)^2 + d \cdot (W - x)^3 + \ldots \tag{3.29b}$$

Because this approximation holds for an arbitrary value of x, set $x = 0$, and consider a *randomly distributed rate of return, R_j*,

$$U(R_j) = a + b \cdot R_j + c \cdot R_j^2 + d \cdot R_j^3 + \ldots \qquad (3.30)$$

Taking the expected value of both sides,

$$E[U(R_j)] = a + b \cdot E(R_j) + c \cdot E(R_j^2) + d \cdot E(R_j^3) + \ldots \qquad (3.31)$$

How many moments of the probability distribution of R_j should be considered? This depends on

- The shape of the utility function (through the parameters a, b, etc.)
- How good of an approximation is desired
- The characteristics of the probability distribution of R_j

If R_j's probability distribution is well approximated by the normal distribution, only the mean and variance of the distribution must be considered. Given the definition of variance, the second moment of the distribution of R_j, $E(R_j^2)$, is equal to $VAR(R_j) + [E(R_j)]^2$. So,

$$E[U(R_j)] = a + b \cdot E(R_j) + c \cdot \{VAR(R_j) + [E(R_j)]^2\}$$

$$= a + b \cdot E(R_j) + c \cdot \{[SD(R_j)]^2 + [E(R_j)]^2\} \qquad (3.32a)$$

i.e., the expected utility associated with an uncertain rate of return is a function of its mean and standard deviation. Setting parameters equal to reasonable values ($b > 0$ and $c < 0$):

$$E[U(R_j)] = a + b \cdot E(R_j) - c \cdot [SD(R_j)]^2 - c \cdot [E(R_j)]^2 \qquad (3.32b)$$

To simplify this exercise, assume the following representation for Equation (3.32b):

$$E[U(R_j)] = [SD(R_j) - a]^2 + [E(R_j) - b]^2 \qquad (3.32c)$$

Let $a = 0$ and $b = 10\%$. Further, define the expected utility function such that $E[U(R_j)] = [1/E(R_j)^2]$, so that lower values of *rate of return variance* represent higher levels of well-being. Plot Equation (3.32c) for $SD(R_j) = 0, 1, 2, 3, 4, 5, 6,$ and 7 for $U = 0.115, 0.125,$ and 0.135. These curves represent an individual's *indifference* between increases in risk and higher rates of return at different levels of utility. (Note: Circular *indifference curves* violate the axioms of utility theory and are used here only for illustration.)

What is the highest level of utility such that the indifference curve is tangent to the line defined in Exercise 3.1, i.e., $E(R) = 0.5 + 0.6025 \cdot SD(R)$? What is the approximate optimal combination of risk and return for this particular individual? At that optimum, what portion of the portfolio should be held in risk-free assets and what remaining portion should be held as a share of the market portfolio?

EXERCISE 3.3. THE CAPITAL ASSET PRICING MODEL

To determine the risk premium that investors seek on risky assets, the ***Capital Asset Pricing Model*** (CAPM) observes the following relationship:

$$E(R_j - R_f) = \beta_j \cdot E(R_m - R_f) \tag{3.33a}$$

where
$E(R_j - R_f)$ is the expected risk premium above the risk-free interest rate for asset j,
$E(R_m - R_f)$ is the expected risk premium above the risk-free rate on a market portfolio (which has been around 8% during the last 70 years in the US), and
β_j is a measure of asset-specific risk (defined below).

Because $E(R_f) = R_f$, since the risk-free asset has no variance,

$$E(R_j) - R_f = \beta_j \cdot E(R_m - R_f) \tag{3.33b}$$

If the investor can invest in a risk-free asset and an efficient market portfolio, the expected return on the resulting portfolio is

$$E(R_j) = R_f + \beta_j \cdot E(R_m - R_f) \tag{3.33c}$$

The expected return, $E(R_j)$, equals the risk-free rate of interest, R_f, plus a risk premium based on (1) the riskiness of the asset, β_j, times (2) the expected risk premium on the market portfolio, $E(R_m - R_f)$. For example, if $R_f = 6\%$ per year and $E(R_m - R_f) = 8\%$, then the expected return on the market portfolio (where $\beta = 1$) would be 14%. If $\beta_j = 1.1$, then $E(R_j) = 14.8\%$.

Under the assumptions of the mean-variance approximation, the CAPM shows that

$$\beta_j = COV(R_j, R_m)/VAR(R_m) \tag{3.34}$$

The risk-free asset has a $\beta = 0$ because there is no covariance between a constant, R_f, and the randomly distributed return on the market portfolio, R_m. Therefore, the expected *risk premium* on the risk-free asset is 0. On the other hand, if $R_m = R_j$, then the right-hand side is only equal to the left-hand side when $\beta = 1$, i.e., the β on the market portfolio is 1. Because investors (in equilibrium) demand the expected *market rate of return* on the *market portfolio*, Equation (3.34) states that the risk of the individual asset is equal to its covariance with the market rate of return normalized by the variance of the market rate of return.

To determine an expected rate of return on the stock of a particular firm (such as an electric utility), we need values for four variables:

1. The risk-free rate of interest
2. The expected risk premium on an efficient market portfolio
3. The covariance of the firm's return with the return on the market portfolio
4. The variance of the return on the market.

3.3.1. Calculating β_j

Calculating β_j is done by "regressing" R_j on R_m using Ordinary Least Squares (OLS):

$$R_j = \alpha_j + \beta_j \cdot R_m \tag{3.33d}$$

where α_j is a measure of "abnormal" returns; its expected value is zero. β_j is equal to its value in Equation (3.34) by the definition of OLS. (Although this version of the CAPM is different from Equation 3.33a, it yields the appropriate value for β_j defined in Equation 3.34.) Using regression routines (for example, in a spreadsheet computer program), estimate α_j and β_j for the stock returns in Table 3.1. Let R_j be the "monthly percentage returns" for each stock. Let R_m be the return on the NYSE. Plot returns of American Express and Chevron against the return on the NYSE. Plot the observations and regression lines for American Express and Chevron using a spreadsheet program.

3.3.2. Calculating *WACC*

The expected return on the assets of a diversified firm can be thought of as the expected return on a portfolio of the firm's profit generating activities. Just as the financial investor can reduce risk by diversifying, the firm can manage the risk and return on its common stock by financing the firm's operations and investments with a combination of at least two financial instruments. If the firm has issued both debt and equity, the nominal *Weighted Average Cost of Capital* (*WACC*) is

$$WACC = (Debt/Value) \cdot R_{debt} + (Equity/Value) \cdot R_{equity} \tag{3.35a}$$

where

Debt is total outstanding debt

Equity is total outstanding equity

Value is *Debt* plus *Equity*

R_{debt} is the nominal rate of interest on the firm's debt obligations

R_{equity} is the nominal expected rate of return on the firm's equity

R_{equity} is determined as

$$R_{equity} = R_f + \beta_{equity} \cdot (R_m - R_f) \tag{3.35b}$$

where β_{equity} (the riskiness of the firm's equity) is given by the CAPM. Calculating the *WACC* for the firm using the CAPM approach requires information on

1. The nominal risk-free rate of interest, R_f; given by the rate of interest on government bonds
2. The firm's nominal cost of debt, R_{debt}, given by the rate of interest on its corporate bonds
3. The nominal expected return on an efficient market portfolio, R_m
4. The capital structure of the firm (*Debt*/*Value* and *Equity*/*Value*)
4. An estimate of the firm's β

Calculate the *WACC* for an electric utility with the following financial characteristics (all values are expressed in nominal rates):

1. $R_f = 6\%$
2. $R_{debt} = 8\%$
3. $R_m = 14\%$

4a. *Debt*/*Value* = 33%

4b. *Equity*/*Value* = 67%

5. $\beta = 0.5$

where β is the approximate value for electric utilities in the US during the 1990s.

EXERCISE 3.4. CERTAINTY EQUIVALENT DISCOUNT RATES

3.4.1. Calculating Risk Premiums

When determining the Present Value (*PV*) of uncertain future returns, there are two related methods: (1) risk-adjusted discount rates and (2) certainty equivalence. Risk-adjusted discount rates were discussed in Section 3.1:

1. The *real (risky)* rate of interest, *r*, is equal to the ***real*** *risk-free* rate plus a *risk premium, RP*:

$$r = r_f + RP \tag{3.36a}$$

2. The *nominal (risky)* rate of interest, *R*, is equal to the *nominal risk-free* rate plus a *risk premium, RP**:

$$R = R_f + RP^* \tag{3.36b}$$

Here, both *r* and *R* are risk-adjusted discount rates. If the riskiness of the project does not change during the project's life, then a single (time invariant) discount rate is appropriate. However, if project risk changes over time, the risk premium must change with changes in risk. Consider

$$PV = C_0 + C_1/(1 + r_1) + C_2/(1 + r_2)^2 \quad (3.37a)$$

where C_0 is the initial investment in the project and C_t is the net cash flow in period t. Substituting $r_t = r_f + RP_t$,

$$PV = C_0 + C_1/(1 + r_f + RP_1) + C_2/(1 + r_f + RP_2)^2 \quad (3.37b)$$

Let $r_f = 5\%$, $RP_1 = RP_2 = 10\%$, $C_0 = -150$, and $C_1 = C_2 = 100$, then

$$\begin{aligned} PV &= -150 + 100/(1 + 0.05 + 0.10) + 100/(1 + 0.05 + 0.10)^2 \quad (3.37c) \\ &= -150 + 100 \cdot 0.87 + 100 \cdot 0.76 \\ &= 13 \end{aligned}$$

However, if much of the uncertainty is resolved after the first year of operation, the risk premium should reflect changes in risk. For example, if $RP_2 = 5\%$, then the PV becomes

$$\begin{aligned} PV &= -150 + 100 \cdot 0.87 + 100/(1 + 0.05 + 0.05)^2 \quad (3.37d) \\ &= -150 + 100 \cdot 0.87 + 100 \cdot 0.83 \\ &= 20 \end{aligned}$$

which is higher than under the single (time-invariant) risk-adjusted rate in Equation (3.37c).

3.4.2. Calculating Certainty Equivalence

To help distinguish between discounting for risk and discounting for time, one approach is to define a cash flow that has certainty equivalence to a risky cash flow. This Certainty Equivalent (CEQ) cash flow can then be discounted *using the risk-free rate*. Considering only one period,

$$PV = C_1/(1 + r) = CEQ_1/(1 + r_f) \quad (3.38a)$$

Given a risk-adjusted discount rate, r, and a risk-free rate, r_f, what value for CEQ solves the second equality? For the example above,

$$PV = 100/(1 + 0.15) = CEQ_1/(1 + 0.5) \quad (3.38b)$$

$$CEQ_1 = 100 \cdot (1 + 0.5)/(1 + 0.15) = 91.30 \quad (3.38c)$$

i.e., an investor is indifferent between receiving a risky \$100 cash flow or a certain

\$91.30 cash flow. Using this same technique, all future cash flows can be adjusted to their certain equivalents, then discounted to the present *using the risk-free rate.*

On the other hand, the investor could determine the perceived certainty equivalence, then calculate the risk-adjusted discount rate that corresponds to this perception. For example, let $CEQ_1 = \$90$, i.e., for a project the investor is indifferent between receiving the risky \$100 return or the certain \$90 return in one year. Solving for the risk premium, *RP*:

$$100/(1 + 0.05 + RP) = 90/(1 + 0.5)$$

$$RP = (100 - 90) \cdot (1 + 0.05)/90 = 11.67\%$$

So, $r = 5\% + 11.67\% = 16.67\%$. Therefore, the investor's implicit risk-adjusted discount rate is greater than the assumed rate of 15%. In this situation, more marginal projects might be accepted than the investor should optimally accept.

3.4.3. Calculating Certainty Equivalence Using the CAPM

One can also determine the risk-adjusted discount rate following the CAPM. Since

$$r = r_f + \beta \cdot (r_m - r_f) \qquad (3.39a)$$

where

$$\beta = \mathrm{COV}(r, r_m)/\mathrm{VAR}\,(r_m) \qquad (3.39b)$$

then

$$r = r_f + \mathrm{COV}(r, r_m)/\mathrm{VAR}\,(r_m) \cdot (r_m - r_f) \qquad (3.39c)$$
$$= r_f + \mathrm{COV}(r, r_m) \cdot [(r_m - r_f)/\mathrm{VAR}\,(r_m)]$$

The last term is known as the "market price of risk" and is often designated as λ. So,

$$r = r_f + \lambda \cdot \mathrm{COV}\,(r, r_m)$$

Therefore, when β is unknown, but the market price of risk (which is independent of r) and $\mathrm{COV}(r, r_m)$ are known, this approach can be used. Here, if $\mathrm{COV}\,(r, r_m) = 0$, then $r = r_f$, i.e., if r is independent of r_m, investors should be willing to accept a return equal to r_f.

3.4.4. Certainty Equivalence of Electricity Generating Investments

Assume that an electric utility would like to invest in a new technology, for example, a new combined-cycle gas turbine power plant (CCGT). The first unit will take a year to construct and cost \$125,000,000. The electric utility believes that there is only a 50% chance that it will produce power at a price lower than competing sources. If power is cheaper, the utility will construct eight more units for \$1 billion and will earn \$250 million *in profit* per year in each year of operation. If the electric power is more expensive than the alternative, the utility will abandon the project and declare a \$125M loss.

The expected cash flows are (assuming that there are no revenues in year 0 or 1)

- $E[C_0] = -\$125M$
- $E[C_1] = 50\% \cdot 0 + 50\% \cdot -\$1B = -\$500M$
- $E[C_t] = 50\% \cdot 0 + 50\% \cdot \$250M = \$125M$, for $t = 2, \ldots, T$

where T is the life of the project. Because of the 50% probability of not building the eight additional units, the electric utility argues that they should apply a high-risk premium to this project. Therefore, the utility discounts these cash flows at 25%.

Assuming $r = 25\%$, the present value of C_1 would be $-\$500M/(1 + 25\%) = -\$400M$. Treating the \$125M, $E[C_t]$, as a perpetuity, the present value of these revenues would be $\$125M/0.25 = \$500M$. Under these assumptions,

$$NPV = E[C_0] + PV(E[C_1]) + PV(E[C_t]) = -\$125M - \$400M + \$500M = -\$25M$$

So the utility would not undertake the project.

However, the electric utility has assumed that the risk of the project is constant over time. *After* the construction of the first unit, much of the risk has been resolved. There is a 50% probability of –\$125M on the project and a 50% probability of building the other eight units and earning \$250M per year on an additional investment of \$1B. These revenues should be discounted at a lower than 25% rate because the uncertainty will have been resolved if the additional units are built. Assume that similar projects earn a 10% rate of return. When the utility must decide to build the additional units, there is a 50% probability that

$$NPV = -\$125M - \$1000M + \$250M/0.10 = \$1375M$$

Therefore, with an initial investment of \$125M the utility has an expected return of

$$E(NPV) = 0.5 \cdot -\$125M + 0.5 \cdot \$1375M = \$625M$$

Of course, the certainty equivalent of an uncertain \$625M is less than \$625M.

Assuming the electric utility feels that the certainty equivalent is worth only 50% of \$625M and the risk-free rate is 6%, determine the NPV for the project, using the following formula:

$$NPV = C_0 + CEQ_1/(1 + r_f)$$

(This exercise is extended in Exercise 4.5. These exercises are based on Rothwell and Sowinski, 1999.)

EXERCISE 3.5. CALCULATING THE INTERNAL RATE OF RETURN

The internal rate of return for an investment in a power plant is determined by solving the following equation (see Equation 3.12c) for *IRR*:

$$FC_0/(TR - VC) = (1/IRR) - [1/IRR \cdot (1 + IRR)^T]$$

where FC_0 are fixed costs in the investment period, *TR* are total revenues in each period, and *VC* are total costs in each period. Assume FC_0 = \$750M and $(TR - VC)$ = \$88.7M. This equation is difficult to solve analytically, so it is usually solved numerically. If at 10%, NPV for the power plant is positive, and at 12%, the NPV for the power plant is negative, $0.10 < IRR < 0.12$. Using numeric methods, solve for *IRR*.

CHAPTER 4

ELECTRICITY REGULATION

4.1. INTRODUCTION TO ECONOMIC REGULATION

Chapter 2 discussed the problems of achieving economic efficiency in natural monopoly industries. In these industries, the largest firms can charge the lowest prices, driving rivals from the market. Once there is no competition, the remaining firm can charge monopoly prices, reducing quantity and social welfare. There are several solutions to this problem, including (1) government ownership of the industry with a mandate to provide adequate output at reasonable prices and (2) private ownership with government regulation to provide adequate output and a reasonable return on private investment. Today, in the electricity sector, only transmission and distribution exhibit natural monopoly characteristics. Generation and retail sale of electricity do not appear to be natural monopolies and can be efficient in a competitive setting. We discuss competitive electricity markets more completely in Chapter 5.

The economic theory of regulation attempts to predict which institutional arrangement is preferable as a function of the comparative social costs and benefits of

- Government monopoly
- Private monopoly without regulation
- Private monopoly with (national, regional, or local) regulation

Each solution involves costs, including the social cost of the monopolist using its market power, the cost of maintaining a regulatory agency, and the costs imposed on the monopolist by the regulator.

For example, under all forms of monopoly, managers and employees might not work as hard as they would under competition. In addition to the administrative costs associated with regulation, another potential cost is associated with trying to correct for market failure: misguided regulatory intervention can create social welfare loss. Therefore, the regulator must carefully consider the costs and benefits of each regulatory requirement on the regulatory agency and the regulated utility.

Chapter 3 discussed the private firm's cost of capital. We know in a competitive industry with free entry and exit that *economic* profit is driven to zero. In these industries, the equilibrium rate of return on capital is equal to the returns in competitive industries with similar levels of risk. This rate of return is considered the normal or *reasonable rate of return*. The role of regulation is (1) to encourage enough investment to meet customer demand and (2) to compensate investors with a reasonable rate of return. The first section of Chapter 4 explores the institutional structures and purposes of regulation. Also see Kahn (1991).

There are several ways of accomplishing regulatory goals in the electric power industry. Sections 4.2 and 4.3 explore two basic regulatory forms for determining the appropriate level of revenues for electricity services providers. These are (1) *Rate-of-Return* (ROR) or *Cost-of-Service* (COS) regulation that requires the regulator to actively monitor the electric utility and (2) *Performance-Based Ratemaking* (PBR). Under ROR or COS regulation, the regulator determines (1) appropriate expenses, (2) the value of invested capital, and (3) the allowed rate of return on invested capital. This process requires a costly exchange of information between the regulator and the electric utility. PBR involves mechanisms that attempt to reduce the cost of regulation. Section 4.4 discusses how to design a price structure that allows utilities to earn their allowed rate of return. The final section of Chapter 4 describes the accounting systems used by the US Federal Energy Regulatory Commission (FERC) to determine reasonable wholesale electricity prices.

4.1.1. Regulatory Policy Variables

To achieve the goals of electricity regulation, regulators employ several policy variables:

- *Regulated tariffs* defining the rate structure (discussed in Section 4.4)
- *Allowed investment* in generation (if it is not deregulated), transmission, and distribution assets
- *Access rules*, including entry into the market by *Nonutility Generators* (NUGs), access to the transmission network for wholesale customers, and access to the distribution network for retail customers
- *Quality of service requirements*, including reliability and voltage disturbances

If the goal of the electricity provider is to maximize profit, the regulator must carefully consider each of these policy variables. For example, focusing on price, quantity, and quality, the electricity supplier might have an incentive to focus on low-cost customers (such as those in urban, concentrated areas) and avoid high-cost customers (such as those in rural, less densely populated areas).

The primary *short-run* focus of the electricity regulator is to set a tariff structure that provides the utility with revenues to pay suppliers and compensate the firm for its investments. Expanding on our definitions in Chapter 2, let *TR*, total revenue, equal the sum of revenues from each customer class:

$$TR = \Sigma P_j \cdot Q_j \tag{4.1}$$

where
P_j is the price or tariff charged to each customer class j
Q_j is the quantity sold to each customer class j

Customer classes can include different categories of customers, such as residential, commercial, industrial, and governmental. Or more generally, classes could include purchases made at different times of day, as under *Peak-Load Pricing*, discussed below and in Boiteux (1960). (More recent discussions can be found under *Time-of-Day Pricing* and *Real-Time Pricing*. See, for example, Berg and Tschirhart, 1988; Crew, Fernando, and Kleindorfer, 1998; and Patrick and Wolak, 1999.)

Further, quantities could be divided into electricity production activities, such as generation, transmission, and distribution. Also, because these *revenues* are *required* by the utility to provide the level of service required by customers, they are known as ***Required Revenues***, *RR*. We will discuss the complexity of the tariff structure in Section 4.4. Until then, we will focus on the determination of total revenues to compensate expenses and investments.

The primary *long-run* focus of the electricity regulator is to encourage the electricity supplier to construct enough generation, transmission, and distribution capacity to "meet all demand." If the supplier is earning a reasonable return on investment, there should be adequate incentive to build new capacity. (However, if the supplier is also earning a return on other inputs, such as fuel, there might be an incentive to use more noncapital inputs than is efficient. For example, if the electricity generator is also earning a return on providing natural gas, the electricity generator might not invest in new generating capacity and instead run existing natural gas plants beyond efficient capacities.)

Also, if the supplier is earning an above-normal rate of return on investment, the utility has an incentive to build more capacity than is necessary. Because of this problem, the regulator generally requires some form of *license* to construct new capacity. (This license is the *Certificate of Need* in some jurisdictions or *Certificate of Public Convenience and Necessity* in others.) This licensing process could involve an informal review of the proposed facilities or a formal hearing procedure to decide whether the new capacity is prudent. Sometimes this involves the regulator in *least-cost* or ***Integrated Resource Planning*** (IRP) to determine whether the technology or system being proposed will result in the lowest required revenues. (Least-cost IRP determines minimum cost expansion plans, i.e., a balanced mix of generating units, and transmission and distribution installations, integrating end-use demand-side actions; see Stoll, 1989. See Hall, 1998, and Hall and Hall, 1994 for examples of IRP. See Marnay and Pickle, 1998 for the application of a specific IRP program, ELFIN, to investment planning.) This involvement increases the cost of regulation, so a cost–benefit calculation must be made before the regulator becomes involved in this activity. Or, this process can be delegated to a different regulator, increasing the checks and balances in the regulatory system. For example, first, price regulation could be delegated to a local regulator who is familiar with local suppliers and customers. Second, capacity regulation could be delegated to a regional or national regulator who can coordinate new investment, provide expertise regarding new technologies, and monitor national and international fuel markets.

Another set of policy instruments (*access rules*) involves entry and exit. Under competition, the actual or potential entry of competing suppliers reduces market power (to increase prices above marginal cost). Electricity regulation involves a contract that gives the electricity provider the exclusive right to distribute electricity in exchange for submission to regulatory control. However, generating electricity can be the by-product of industrial processes (cogeneration) and it would be socially inefficient not to distribute possible excess electricity from these industries. Therefore, the regulator must decide who can sell electricity to the exclusive distributor and at what price. These NUGs, (or *Independent Power Producers*, IPPs) can be subject to the same case-by-case rate hearing as the monopoly distributor but, usually, the regulator makes a general ("substantive") rule determining the price paid by the distributor to the NUG. In the US during the last two decades, much controversy has surrounded the determination of ***avoided costs*** (of the monopoly utility) paid for NUG electricity. We will not discuss how various jurisdictions have calculated avoided costs because the introduction of competition into electricity generation avoids this issue.

Regarding exit, this problem does not present itself in the same way as firms exiting from competitive markets, unless, of course, the firm declares bankruptcy and the regulator must find a new owner of the monopolist's assets and franchise. More generally, exit involves attempts by electric utilities to abandon classes of high-cost customers, for example, those living in sparsely populated areas. Once customers are served by the transmission and distribution network, the attempt to abandon them usually disappears, but the electric utility might focus resources to serve low-cost customers, and the regulator might need to persuade the utility to increase capacity in high-cost areas. If capacity in these areas does not meet demand, the regulator could provide financial incentives for other providers to enter these markets, for example, by granting franchises to local cooperative distributors.

Finally, although it is difficult to measure quality in many markets, quality is generally defined in electricity markets as (1) reliability, (2) power quality associated with power outages and voltage disturbances, and (3) customer satisfaction with service. The case studies in Chapters 6–9 explore some *minimum standards* that regulators have attempted to maintain in their jurisdictions and the penalties that are imposed if these standards are not maintained.

4.1.2. The Regulatory Process

As discussed above, the economic theory of regulation maintains that the institutional arrangement that eventually is preferable in a regulated industry is the one that maximizes social welfare through minimizing social costs and maximizing social benefits. In some countries, it is possible that government ownership with government oversight maximizes welfare. However, in many countries, private ownership with independent regulation is better able to minimize the social costs of providing electricity. Below, we discuss the U.S. approach to regulation.

The transition from one system to another (or for a change in the regulatory system) usually begins with an act of the *legislative* branch. It falls to the *executive* branch to make the change and to the *judicial* branch to oversee that change conforms to the act of the legislature. This balance of rights and responsibilities is reproduced in the regulatory system, where the economic power of the *electric utility* is balanced by the regulatory power of the *regulating agency* and overseen by a *judicial* body that also resolves disputes. Most regulatory agencies are headed by a commissioner or commission that is ultimately responsible for the agency's decisions. However, the regulatory staff gathers much of the information and makes recommendations to the head of the agency. Therefore, there is a balance of power between the regulatory staff and the utility with the head of the agency acting as a judge whose decisions can be appealed to a higher judicial authority. Although different nations follow different systems, most rely on a system of checks and balances to ensure that one party in the regulatory process is unable to dominate the others.

Because of the importance of balancing powers between the utility and regulator it is essential to ensure independence

1. Between the regulator and other branches of government to reduce the political influence in setting tariffs and allowing entry into potentially competitive industries
2. Between the regulator and the utility to reduce the possibility of maximizing utility profits at the expense of consumers

The importance of independence is so great that many national governments also enact *conflict of interest* laws. For example, utilities cannot hire commissioners or members of the regulatory staff immediately after their service at the regulatory agency.

Regulatory proceedings can take on the character of a judicial hearing. For example, in the state of California (see Chapter 6), the California Public Utilities Commission makes all decisions acting as a whole. But because of the complexity of regulating many industries, hearings are usually conducted by appointed Administrative Law Judges, who make recommendations to the Commission based on evidence presented by the regulatory staff, the utility, and other interested parties. Although any party with standing can appeal these decisions to the California Supreme Court, the Court rarely overturns Commission decisions. However, consumers affected by the decision can ask the legislature to pass new legislation or circulate a petition to change decisions through a voter referendum.

This judicial orientation leads to case-by-case decisions, although the regulatory agency can make substantive rules that apply to all regulated firms. All these proceedings take time and involve social cost. To minimize the social cost of regulation, the regulated firm has the right of *due process*, i.e., the regulatory agency and the judicial system must meet judicial deadlines following prescribed procedures or forfeit the specific powers of regulation, such as setting price. On the other hand, because utilities can strategically manipulate the regulatory process (by delaying the provision of information to the regulator), utilities must also comply with deadlines and procedures or face penalties.

This regulatory process of balancing powers has developed in the electricity industry into ROR or COS regulation under the assumption of natural monopoly. We discuss traditional forms of regulation in this chapter. But these forms of regulation are not perfect and there have been recent experiments to improve them by introducing competitive markets in generation and retail, as discussed in Chapters 5–9. (Transmission and distribution will likely continue to be regulated under the forms of regulation discussed in this chapter.)

4.2. RATE-OF-RETURN REGULATION

Rate-of-Return (ROR), also known as ***Cost-of-Service*** (COS), regulation involves two steps during the *rate case*. (Much of this discussion is based on the implementation of ROR in the US; other countries employ a slightly different implementation.) First, the *rate level* determination is concerned with (1) identifying allowed costs and investments and (2) setting an ***allowed rate of return*** so that the utility will have the appropriate level of earnings on its investment. During the rate case, tariffs are set based on a test period (generally, the previous accounting period) and remain in effect until the next rate case. Second, the ***rate structure*** determination deals with setting tariffs for different customer classes and products that permit the utility to recover the revenues required to earn its allowed rate of return. Sections 4.2 and 4.3 discuss the rate level and Section 4.4 discusses the rate structure.

Both the utility and the regulator have the right to request a rate case with the following structure.

1. The utility (or the regulatory staff) argues that current tariffs are too low (or high) because the cost allowance or the allowed rate of return is too low (or high).
2. After presentation of detailed accounting information by the utility and the regulatory staff and negotiation between the regulator, agency staff, and the utility, the regulator determines the appropriate level of expenses and sets the allowed rate of return.
3. Tariffs are adjusted to yield the new rate of return allowed by the regulator.

(Changing tariffs also changes quantities, so the regulator also needs information on demand elasticities and the anticipated quantity demanded.)

The following accounting equation summarizes the process of determining the rate level.

$$RR = Expenses + s \cdot RB \tag{4.2}$$

where

RR is required revenues

Expenses is allowed expenses
s is the allowed rate of return
RB is the rate base, the value of the *allowed* investment.

The underlying idea is that a utility's revenues must equal its costs, so that *economic* profit is zero. Notice that *actual* costs can be different from *allowed* costs. (On the difference between allowed and realized rates of return, see Rothwell and Eastman, 1987.) This provides an incentive to the utility for efficient behavior or a penalty for inefficient practices. Table 4.1 shows the accounting components used to calculate the allowed rate of return, s, in Equation (4.2). An adjustment is made to increase revenues and thus the allowed rate of return.

After the rate case, and until the next rate case, required revenues (per MWh) remain fixed. This provides an incentive for the utility to reduce cost. The utility earns higher rates of return by incurring lower costs than the costs anticipated in the rate case. Of course, if costs are higher than anticipated, the utility earns less than the allowed rate of return. The utility's incentive to reduce cost increases as the ***regulatory lag*** increases. (Regulatory lag is the period between two consecutive rate cases, or the lag between a change in required revenues and the next rate case.) If the regulator could adjust tariffs to keep the rate of return always equal to s, there would be no regulatory lag and, thus, no incentive for cost reduction.

Another important instrument that provides incentives for regulated firms to be efficient is the determination of whether a particular investment is to be included in the rate base. The evaluation of the rate base can be done using different methods. These include the following:

1. *Original cost valuation*—the amount that the company originally paid for its plant and equipment, *less depreciation*.

Table 4.1. Overview of Financial Statements

	Anticipated	Adjustment	Anticipated after Adjustment
Revenues	30,000,000	1,600,000	31,600,000
Expenses			
Fuel	24,000,000		24,000,000
Operations	3,000,000		3,000,000
Depreciation	1,000,000		1,000,000
Total expenses	28,000,000		28,000,000
Net operating income	2,000,000		3,600,000
Rate base (*RB*)			
Plant minus depreciation	42,000,000		42,000,000
Working capital	350,000		350,000
Total rate base	42,350,000		42,350,000
Rate of return	4.72%		8.50%

Source: Adapted from Viscusi, Vernon, and Harrington (2000, p. 363).

2. *Reproduction cost*—an estimate of the current cost of reconstructing the same plant and equipment.
3. *Replacement cost*—what it would cost to replace plant and equipment with capacity embodying the newest technology.
4. *Market-value*—the value that might be received if the firm or its assets were sold.

Presently, many regulators in the US use *original cost valuation* and focus on the selection of the allowed rate of return.

The most common method of choosing the allowed rate of return is to calculate the weighted (average) cost of the utility's financial securities, for example, bonds and stocks. (See Chapter 3 on the weighted cost of capital.) Table 4.2 presents an example. Usually, the most controversial issue is to determine the *cost of equity capital*, which depends on a comparison of the utility with similar utilities.

Throughout the rate case, there is an ***asymmetric information*** problem. This results from an asymmetry of information between a regulator who wants something done and the electric utility that must do the work. One solution to this problem is to involve regulators in managerial decisions. For instance, regulators could be involved in the investment planning process, with access to detailed information on utility short-term and long-term costs and assets. This increases information requirements and the costs of regulation.

Another problem with ROR regulation is the "Averch–Johnson effect" (Averch and Johnson, 1962). A firm with an allowed rate of return greater than its cost of capital ($s > r$) has incentives to use too much capital, resulting in inefficiently high production costs. But if its allowed rate of return is less than its cost of capital, it has an incentive to use too little capital and too much of its variable inputs, resulting in inefficiently high production costs. Averch and Johnson assume that the utility maximizes profit subject to the rate-of-return constraint:

$$PR = TR(L, K) - w \cdot L - r \cdot K \tag{4.3}$$

subject to

$$s = (TR - w \cdot L)/K$$

where r is the utility's cost of capital and s is the allowed rate of return. Using the

Table 4.2. Example Calculation of the Allowed Rate of Return

	Percent of Capitalization	Rate (%)
Bonds	70	7.00
Common stocks	30	12.00
Total	100	8.50

techniques of constrained optimization, Averch and Johnson show that the utility has an incentive to use more capital than is efficient as the allowed rate of return rises above the cost of capital. See Exercise 4.4. This inefficiency increases with increases in this divergence. Although it is easy to show this result in a simple model of ROR regulation, it is difficult to prove empirically that the Averch–Johnson effect has had a large influence on utility behavior. Whether or not there is an Averch–Johnson effect, Averch and Johnson's analysis emphasizes the importance of correctly assessing the utility's cost of capital.

In contrast, incentive regulation (discussed next) provides incentives for cost reduction through increasing the regulatory lag between rate cases; therefore, it reduces the costs of regulation incurred by the regulatory agency and the distortions due to the regulatory structure (Joskow and Schmalensee, 1986).

4.3. PERFORMANCE-BASED RATEMAKING

Performance-Based Ratemaking (PBR) or *incentive regulation* is a form of utility regulation that strengthens the financial incentives to lower rates or lower costs, or to improve nonprice performance compared with traditional ROR regulation (Navarro, 1996). PBR weakens the link between a utility's regulated tariffs and its costs. This decoupling is accomplished by decreasing the frequency of rate cases and increasing the regulatory lag. During a regulatory period of several years, e.g., four or five, the regulator establishes a formula that sets utility's revenues or tariffs. After that period, a complete revision of costs and investments (similar to a rate case, described above) takes place, and a new revenue or tariff formula is established for the next regulatory period. The design and application of a PBR plan is a set of interrelated tasks:

1. *Set a baseline revenue requirement or "starting point."* During the rate case, the regulator sets the initial revenues that will be adjusted during subsequent years. The regulator could begin, for example, with a detailed analysis of each cost item or, where comparable "efficient" companies exist, with a benchmark model. The regulator can use benchmarking with engineering and management analysis. (See, for example, Chapter 7, Section 7.5.)
2. *Set the adjustment factors.* Adjustment factors determine how revenues evolve until the end of the regulatory period. These factors could consider variations in macroeconomic indices or changes in the number of customers that might modify the company's efficient costs.
3. *Design of control mechanisms to meet specific objectives.* Regulators also design control mechanisms to ensure that the utility meets specific objectives associated with (1) energy, industry, or social policy, and (2) quality of supply, security of supply, universal service, environmental regulation, research and development programs, etc. But in attempting to reduce cost, the utility might ignore some of these objectives. In some restructuring experiences, economic penalties have been proposed if the PBR-regulated distribution

utility reduces reliability, customer service, or employee safety below specific limits (see Chapters 6–9 for details).

In recent years, interest in PBR has increased because of economic and technological trends leading to more competition in the electricity industry. (For a review of PBR applied to a specific generating technology, see Che and Rothwell, 1995.) Incentive regulation can be seen as a bridge between traditional regulation and deregulation. The most typical forms of incentive regulation for electric utilities are *sliding scale, price cap*, and *revenue cap*. We discuss each of these below.

First, to show the bridge between traditional and incentive regulation, Laffont and Tirole (1993) present a simple model of incentive regulation:

$$TR = a + b \cdot Costs \tag{4.4}$$

where

TR is actual (*ex post*) total revenues received
a is fixed payment, set *ex ante*
b is sharing fraction, $0 < b < 1$, set *ex ante*
Costs are *ex post* costs

Equation (4.4) shows a relationship between *ex post* revenues and costs based on two parameters set *ex ante*, *a* and *b*. The firm's incentive to minimize costs is inversely proportional to the magnitude of the sharing fraction, *b*. ROR regulation with frequent rate cases can be thought of as *low-powered* incentive regulation with $b = 1$, because the utility is allowed to collect all the incurred costs. Forms of *high-powered* incentive regulation (where *b* approaches 0) are (1) rate freezes, (2) ROR regulation with infrequent rate cases, and (3) PBR revenue or price cap formulas with a long regulatory lag. They increase the portion of revenue a utility receives *ex ante* and decrease the portion of utility revenue computed as a function of *ex post* costs.

A purely competitive market (where the seller cannot influence the market price) is another situation of high-powered incentives. High-powered incentive regulation with no adaptation to *ex post* costs will eventually result in overall revenues that are either too high or too low, and can threaten the viability of the incentive plan. Therefore, a rate case after several years is required. The incentive plan should balance its short-run incentive power, its overall economic efficiency, and its ability to remain viable over time (Comnes, Stoft, Greene, and Hill, 1995).

4.3.1. Sliding Scale

One method of incentive regulation, known as *sliding scale*, is described in Viscusi, Vernon, and Harrington (2000). Its essential property is that it permits the sharing of risks and rewards between the utility and consumers. Under this mechanism, a sliding scale would adjust prices in the current rate case, so that the allowed rate of return, *s*, at the new prices would be

$$s = R_t + h(R^* - R_t) \tag{4.5}$$

where

h is a constant between 0 and 1

R_t is the realized rate of return at the tariffs set in the previous rate case in year t

R^* is the target rate of return

If $h = 1$, tariffs are always adjusted to give the firm a rate of return of R^*. If rate cases were frequent, the firm would neither benefit from being efficient, nor be hurt by being inefficient. Similarly, if $h = 0$, the regulation is *fixed-price*: All gains from efficiency accrue to the firm, and all unexpected cost increases affect the firm alone. An h value of 0.5, however, would indicate that unexpected benefits and costs are shared between the firm and its customers.

In Comnes, Stoft, Greene, and Hill (1995), *sliding scale* regulation is presented as a mechanism under which tariffs are adjusted to keep a utility's rate of return within (or close to) a prespecified ROR band. The primary rationale is to reduce the frequency of rate cases and thus increase regulatory lag. If earnings become too high above the band, rates are cut; if earnings fall too low below the band, rates are increased. Adjustments outside the band are sometimes only partial. Therefore, sliding scale is an *earning–loss sharing mechanism* between the utility and its customers. If earnings fall within the specified band, rates are not changed, and the firm makes profits or losses.

Earnings-sharing mechanisms are sometimes used in combination with *price caps* or *revenue caps*; see below. A progressive sharing mechanism is one in which the utility's share of the cost savings increases with the cost savings achieved. For instance, the utility might receive 20% of the first 1% of its cost savings, 40% of the next 1%, and so on. Because each increment of additional cost savings will usually cost the utility more to achieve, progressive mechanisms are preferred to regressive mechanisms in which the utility's share falls as cost savings rise. (See Exercise 4.1.)

4.3.2. Revenue Caps

Under ***revenue cap*** regulation, the utility's allowed revenues are set during a regulatory period of several years, usually four or five. The revenue amount allowed during the first year is adjusted in subsequent years according to a prespecified set of economic indices and factors. Subject to this cap, the utility is permitted to maximize its profit during the regulatory period by minimizing total costs. When the regulatory period expires, a rate case takes place, and a new revenue cap formula is set for the following period. A common form of a revenue cap is as follows (Comnes, Stoft, Greene, and Hill, 1995).

$$TR_t = \{[TR_{t-1} + (CGA \cdot \Delta Customers)] \cdot (1 + i - X)\} \pm Z \tag{4.6}$$

where

TR_t is authorized utility revenues in year t

CGA is customer growth adjustment factor (U.S.$/customer)
$\Delta Customers$ is annual change in the number of customers
i is annual change in prices (the inflation index, e.g., CPI or RPI)
X is productivity offset
Z is adjustments for unforeseen events

The objective of the revenue adjustments year-by-year in Equation (4.6) is to simulate how the costs of the company would reasonably change with changes in system features (such as the number of customers or the price of inputs). To measure inflation the use of CPI or RPI is optional, since other price indices (e.g., sector-specific industrial price indices) can also be used. Different values of *Customer Growth Adjustment* (CGA) factors can be specified for different classes of customers. For instance, in case of network distribution revenues, customers connected in medium-voltage networks can be considered with a different adjustment factor than customers connected in low-voltage networks. The X productivity offset factor ensures that customers receive a share of the expected enhanced productivity required of the utility. The unforeseen events ($\pm Z$) can include, for example, increased taxes, changes in environmental laws, natural disasters, or restructuring costs. However, $\pm Z$ should not cover costs that cannot be projected with a reasonable degree of accuracy. In these cases, revenues are adjusted *ex post* according to realized values.

A variant of the revenue cap is the *Revenue-per-Customer* (RPC) cap. Equation (4.6) is a generalized revenue cap. If CGA is equal to the average revenues per customer, then it is equivalent to an RPC cap. Revenue caps do not address retail prices. Therefore, revenue caps usually coexist with traditional methods of allocating costs and setting relative prices for customer classes.

4.3.3. Price Caps

Under ***price caps***, maximum (but not required) prices for utility services are set for several years without regard to the utility's own costs. Maximum prices allowed during the first year are adjusted in subsequent years according to a prespecified set of economic indices and factors. The most common form of price cap is

$$P_{j,t} = [P_{j,t-1} \cdot (1 + i - X)] \pm Z \tag{4.7}$$

where

$P_{j,t}$ is maximum price or tariff that can be charged to the jth customer class in year t
i is annual change in prices (the inflation index)
X is productivity offset
Z is adjustments for unforeseen events

Equation (4.7), known in the UK as the "RPI minus X" and in the US as the "CPI minus X," has been widely applied in incentive regulation of the telecommunications industry in the U.S. and for regulation of electricity distribution utilities in the UK. ROR regulation with rate freezes and significant regulatory lag can be viewed as price cap regulation without adjustment factors. Price caps can be applied to customers as a whole or to individual classes of customers. The number of caps presents a choice to regulators between (1) utility pricing flexibility and (2) preventing cost shifting between customer classes (Woolf and Michals, 1995).

Regulators can combine price caps and revenue caps with profit-and-loss-sharing mechanisms intended to protect both the utility and the ratepayers from the risk of over- or underrecovery of revenues. These mechanisms take effect if the company earns above or below a specified band around its allowed rate of return.

Although revenue and price caps create similar incentives to minimize costs, they differ significantly in the incentives they provide for incremental sales. Price caps create an incentive to maximize sales. On the other hand, revenue caps create incentives to minimize sales (either by raising prices or by other means) to reduce costs. Finally, revenue caps can be considered friendlier to energy savings and ***Demand Side Management*** (DSM) programs. (These are programs that reduce electricity demand; see Comnes, Stoft, Greene, and Hill, 1995; Rothwell, Sowinski, and Shirey, 1995; and Woolf and Michals, 1995).

4.3.4. Some Problems with Incentive Regulation

Concerns with PBR regulation include the following:

1. *Quality of service degradation.* Because the regulated firm is the only provider of services, most customers will continue to buy at the regulated tariff even if quality suffers from cutting costs. A practical implication is that regulators must set standards for quality, monitor utility performance, and penalize poor quality.
2. *Concerns over excessive or low profits cause a convergence with ROR regulation.* As the time between rate cases increases, the utility captures more benefits from any productivity-improving initiatives. Unfortunately, several factors, including considerations to set equitable rates, unforeseen events that influence utility costs, and asymmetric information problems, could increase the rate case frequency, making incentive regulation more closely resemble traditional (ROR) regulation.
3. *Shift of costs toward the most captive customers.* The utility has incentive to shift costs from the unregulated to the regulated lines of business, because under incentive regulation, regulators are relieved of the need to monitor costs. Further, if the PBR plan allows pricing flexibility, it will likely lead to an increase in relative rates for customer classes that have the fewest alternatives.

4.4. RATE STRUCTURE

4.4.1. Introduction to Tariff Regulation

Sections 4.2 and 4.3 discussed how the regulator determines the total revenues required to provide electricity to all classes of customers. This section discusses how to determine tariffs (or prices) for each customer class. Ideally, economists would recommend setting tariffs for each customer at the marginal cost of providing electricity. If the marginal cost of providing an industrial customer is less than the marginal cost of delivering electricity to a household, then households should pay a higher tariff. Unfortunately, different tariffs for different customers, i.e., ***price discrimination***, though economically efficient, can be viewed as inequitable. Also, charging different prices could also involve ***cross-subsidization*** (i.e., charging one customer class a higher tariff so that another customer class can pay a lower tariff). Before discussing pricing strategies for allocating costs among customer classes, we present a brief overview of price discrimination and cross-subsidization.

In Chapter 2, we assumed that the monopolist could not discriminate between customers and could charge only one price. However, if the monopolist can price discriminate, what price would it charge to each customer to maximize profit? If the monopolist (1) knew the maximum price each customer was willing to pay (or the quantities that each customer would buy at each price) and (2) could separate customers so that they could not resell electricity to each other, then the monopolist could negotiate a maximum price for each customer, thus capturing all consumer surplus as profit. This is *first-degree price discrimination*. However, it is difficult to know the maximum willingness to pay.

Second-degree price discrimination is easier to implement and is often used in electricity pricing by price regulators. Under it, the utility charges different tariffs for different amounts of electricity. If marginal cost falls for larger levels of consumption and the willingness to pay declines with larger levels, lower tariffs are charged for larger amounts. For example, if P is the tariff for the first 200 kWh, then $0.8 \cdot P$ could be charged for the next 200 kWh, and $0.6 \cdot P$ could be charged for the next 200 kWh, etc. This is *declining-block pricing*. (This was a common practice for residential pricing in the United States when generation was assumed to be a natural monopoly, but was replaced with *increasing-block pricing* when technological constraints led to increasing marginal costs.)

Also, we define *cross-subsidization*. This is a practice by which some customers are charged a higher tariff, so that other customers can be charged a lower tariff. For example, if rural customers are charged the same tariff as urban customers, but the cost of supplying rural customers (because of the cost of distribution) is higher than the cost of supplying urban customers, then urban customers are *cross-subsidizing* rural customers. There are two commonly used tests to determine the existence of cross-subsidization:

1. The *stand-alone average cost* test
2. The *average incremental cost* test

Under the *stand-alone average cost* test, if the tariff charged to each customer class is less than the average cost would be if each class generated its own electricity, then the price structure does not subsidize one class at the expense of another. Although one class might be paying above marginal cost, it is better off because the average cost of supplying the whole market is lower. However, it is difficult to determine what the average cost would be for each customer class. Under the *average incremental cost* test, if each customer class pays an amount that at least covers the incremental cost of supplying it with electricity, then the price structure does not involve cross-subsidization (see Viscusi, Vernon, and Harrington, 2000).

4.4.2. Marginal Cost Pricing, Multipart Tariffs, and Peak-Load Pricing

Chapter 2 discussed the primary problem with marginal cost pricing in the context of electricity. For a natural monopolist, marginal cost is less than average cost because of large fixed costs. If price equals marginal cost, the utility has a loss on each kWh. There are several solutions. First, prices could be set at average cost, as discussed in Chapter 2. Second, the government could subsidize the utility by making up the difference between marginal cost and average cost. (However, the government must raise taxes to cover this subsidy, and those taxes could introduce other distortions in the economy, such as tax-avoidance behavior.)

Third, the regulator could use a *Ramsey pricing* mechanism (Ramsey 1927). Under this pricing rule, customers or customer classes are charged different tariffs based on their demand elasticity. This is a tariff structure in which tariffs are raised differentially above marginal cost until total revenues equal total costs, i.e.,

$$(P_j - MC)/P_j = c/E_{dj} \tag{4.8}$$

where

P_j is the price or tariff for the *j*th customer or class
c is a constant that depends on the cost structure
E_{dj} is the demand elasticity of the *j*th customer or class

The markup above marginal cost is inversely related to the demand elasticity. Customers with few alternatives to purchasing electricity (such as residential and commercial customers) would be charged a higher tariff than customers with more alternatives (such as industrial customers who could purchase fuel and generate their own electricity). Of course, whenever different tariffs are charged for electricity to different customers, there is an incentive for low-priced customers to sell electricity to high-priced customers, thus avoiding the price discrimination mechanism. In situations in which price discrimination is easy to enforce, such as in transportation services, Ramsey pricing is routinely used to differentiate between the type of product being shipped. *Value-of-service pricing* charges a higher price to manufactured goods (where the price of transportation is low compared with the price of the product) and a lower price to agricultural and mineral products.

Fourth, the regulator can use a ***two-part tariff*** to allocate *fixed costs* among customers and use marginal-cost pricing to allocate *variable costs*. Although this approach has the benefit of signaling the correct price to the consumer, determining how to allocate fixed costs is difficult. If fixed costs (*FC*) are allocated equally and there are *n* customers, then under an equitable distribution, all customers would pay *FC/n*. Customers pay *FC/n* to "enter" ("connect to") the market, then pay *MC* to purchase their desired quantity. This type of pricing mechanism is common among telephone services that charge a high fixed fee to have a telephone installed and a low fee for using the telephone. The high fixed fee covers (1) the cost of connecting the user to the network, including the telephone, and (2) a portion of the cost of providing the telephone network (see Exercise 4.2).

However, in electricity service (and elsewhere) two-part tariffs might exclude some customers who would be willing to pay a marginal cost for each kWh, but who might not be able to pay the connection charge. Of course, the regulator could price discriminate among customers by charging different connection fees. If price discrimination is not politically feasible, the entry fee could be lowered to provide universal service. However, to cover the fixed costs, this would require raising the price above marginal cost for each kWh, causing inefficiencies.

The two-part tariff solution can be extended to a multipart tariff in electricity pricing. For example, there are four major components of the *marginal cost* of electric service:

1. Marginal energy and generation capacity costs
2. Marginal costs of transmission and distribution facilities
3. Marginal energy cost of transmission and distribution networks
4. Marginal customer costs

Investment in transmission facilities depends on the level of peak demand. This means that peak demand should be charged for the cost of expanding transmission capacity. Distribution facilities investments, however, depend on the level of demand assumed when the system was built, so it does not change with increments in actual demand until capacity is surpassed.

These pricing mechanisms can be applied to *fully distributed cost* (FDC) pricing. Under FDC, the regulator (1) allocates costs to serve a particular customer to that customer and (2) divides *common costs* among customers. For example, the costs of connecting customers to the distribution network are assigned to the connected customer. The costs of generation and transmission are common costs that must be allocated among customers. Common costs can be treated as fixed costs in a two-part tariff. Unfortunately, this approach does not necessarily provide the correct marginal cost signals.

Finally, a special case of two customer classes in which cross-subsidization might be present is the case of on-peak and off-peak customers. The problem of setting different tariffs for *base-load* and *peak-load* power has been discussed in the economics literature for many year (see Exercise 4.3). The problem arises because it is difficult without *time-of-day* consumption information to charge

- A *low* tariff for base-load electricity use, such as for refrigeration for residential customers and for street lighting for governmental customers
- A *higher* tariff for peak-load uses, such as evening use for residential customers or on particularly hot or cold days during the year

If one price is charged, then

- Price is greater than marginal cost for base-load customers, so too little is (efficiently) consumed
- Price is lower than marginal cost for peak-load customers, so too much is (efficiently) consumed

This leads to underinvestment in base-load capacity and overinvestment in peak-load capacity.

One solution is to implement ***peak-load pricing***, i.e., different prices for on- and off-peak electricity. However, implementing peak-load pricing based on time-of-day consumption usually requires new metering equipment. Implementing time-of-year pricing is much easier. This could involve charging different tariffs each day, announced the previous day. Although this approach is more efficient than seasonal pricing, it might not be feasible if consumers do not have easy access to daily price information.

Because it might not be efficient to require small consumers to invest in demand-metering equipment or time-of-day energy meters, the tariff elements will vary according to the meters in place. Thus, customers with time-of-day meters should (in theory) face four elements in their tariff (in practice some simplifications could be adopted according to customer sophistication, ease of administration, etc.):

1. An energy charge for kWh consumed in each pricing period, to reflect the marginal energy and generation capacity costs, corresponding to the market price
2. A capacity charge for peak kW used in each pricing period, to reflect the marginal costs of transmission and distribution facilities
3. A fixed charge related to the customer's demand, to reflect the marginal cost of distribution facilities between the substation and the customer's meter
4. A fixed monthly charge, varying according to the customer class, but independent of usage, paid to cover the marginal costs of metering and billing

Customers who do not have meters to measure demand at different times of day would face a similar set of tariff elements, except that (1) and (2) would be combined into a single seasonal charge per kWh.

4.5. OVERVIEW OF THE UNIFORM SYSTEM OF ACCOUNTS

Traditional ROR regulation requires extensive electric utility accounting information. A uniform system of accounts applied to all utilities in a jurisdiction has been under discussion in the accounting profession since the introduction of electric lighting in the late nineteenth century. This system is an extension of Table 4.1 to cover all forms of utility expenses and investments. Associated with the system is an extensive set of definitions describing what is included and excluded from each cost category. Because it is primarily used to determine the appropriate rate of return, these accounts differ from those required by the tax authorities or financial institutions. Therefore, great care has been taken by accountants to design a system that is transparent to the utility's management, the rate regulator, the tax collector, and investors.

The creation of a *U.S. Uniform Systems of Accounts*, and codified in the *U.S. Code of Federal Regulations* (CFR), was the responsibility of the Federal Power Commission (FPC, the precursor of the Federal Energy Regulatory Commission, FERC) because of its role in regulating wholesale and interstate sales of electricity in the US. However, because each state is responsible for regulating retail prices, the National Association of Regulatory Utility Commissioners (NARUC), representing the electric utility regulators, has worked extensively with the FPC and FERC to update the uniform system. Although it is not possible here to adequately describe this system of accounts in the US (and because it is being continually updated), information is available in the public domain and on the Internet. Readers can download information on how these accounts are defined and used in FERC Form 1, "Annual Report of Major Electric Utilities, Licensees, and Others" at www.ferc.fed.us/electric/electric_USOA/electric_USOA.htm.

EXERCISE 4.1. A PROFIT-SHARING MECHANISM UNDER PBR REGULATION

A distribution company presents a financial statement (Table 4.3) to the regulator for the next regulatory period. The regulator has established a profit-sharing mechanism between the shareholders and the ratepayers. This mechanism works as shown in Table 4.4. For example, under a "progressive" mechanism, if the utility reduces

Table 4.3. Anticipated Financial Statement

	Anticipated (1,000$)
Revenues	32,000
Expenses	
Operation and management cost	14,200
Purchase transmission services	7,100
Value of lost electricity	2,556
Depreciation plus taxes	4,554
Total expenses	28,400
Net operating income	3,600
Rate base	42,350
Rate of return	8.50%

Table 4.4. "Progressive" Profit-Sharing Mechanism

Basis points	0–50	50–100	100–150	150–200	>200
Shareholders %	20	40	60	80	100
Ratepayers %	80	60	40	20	0

cost by 1% (= 100 basis points), the shareholders would receive 20% of the first 50 basis points, 40% of the next 50 basis points, and so on.

4.1.1. If the utility decreases its operation and management costs by 4%, determine how much of these savings will be kept by the utility as profits and how much will be returned to ratepayers as a rate decrease.

4.1.2. In addition to this operation and management cost reduction, the utility hopes to reduce electricity losses. In reducing these losses, each increment of loss reduction will cost the utility more to achieve (due to increasing marginal costs). Table 4.5 shows loss reductions and the increment of operational and investment costs needed to achieve each reduction. Determine the optimal utility loss reduction target.

4.1.3. If the regressive profit-sharing mechanism presented in Table 4.6 is implemented instead of the progressive one proposed previously, determine the new optimal cost reduction target.

Table 4.5. Loss Reduction and Cost Increment

Reduction in losses (%)	0–10	10–20	20–30	30–40	40–50	50–60
Cost increment needed (%)	0–1	1–4	4–9	9–16	16–25	25–36

Table 4.6. "Regressive" Profit-Sharing Mechanism

Basis points	0–50	50–100	100–150	150–200	>200
Shareholders %	80	60	40	20	0
Ratepayers %	20	40	60	80	100

EXERCISE 4.2. OPTIMAL TWO-PART TARIFFS

An electric utility has the following cost structure:

$$TC = 500 + 20 \cdot Q$$

Market demand for its electricity is

$$P = 100 - Q$$

4.2.1. If price is set at marginal cost, what is the electric utility's profit?

4.2.2. If price is set at average cost, what are the equilibrium price and output, and what is the deadweight loss compared to Exercise 4.2.1?

4.2.3. Under a two-part tariff, each customer pays a connection charge (fixed fee) to use the electric system and a usage fee for each kWh. If the usage fee is set to marginal cost and there are 10 identical customers, what is the largest connection charge that a customer would be willing to pay? What charge would let the electric utility cover its costs? What would be the deadweight loss?

4.2.4. Next assume that there are six Class 1 customers with demand curves

$$P = 100 - 6.3 \cdot q_1$$

and four Class 2 customers with demand curves

$$P = 100 - 80 \cdot q_2$$

What is the largest connection charge that a Class 2 customer would be willing to pay if the usage fee is set at marginal cost? If the Class 2 customers would not be willing to pay 50 to connect (i.e., the Class 2 customers would not enter the market), the electric utility could cover fixed costs by charging all of them to the Class 1 customers. What would be the resulting deadweight loss? (The market demand curves are determined by adding individual demand curves at each price.)

4.2.5. If the regulator allowed the electric utility to charge a different connection charge for each customer class, what are two feasible connection charges and their associated deadweight losses?

4.2.6. An optimal two-part tariff minimizes deadweight loss by (1) charging an optimal connect charge to all customers equal to the consumer surplus of the Class 2 customer and (2) charging all customers the same usage fee. Determine the optimal two-part tariff using these criteria. (This problem relies on Viscusi, Vernon, and Harrington, 2000, p. 360.)

EXERCISE 4.3. THE PEAK-LOAD PRICING PROBLEM

Assume that a cogeneration heating distribution system can be constructed for \$14,600 per unit of capacity and can be operated at variable costs of \$0 (heat is supplied by an industrial facility that operates 24 hours every day). Assume that during 8 hours of the day the on-peak demand for heat is

$$P = a - b \cdot Q$$

Assume that during the other 16 hours of the day (off-peak) the demand for heat is one-half the demand during the peak period. Assume that the cost of capital, r, is 10%, and that heating distribution system does not depreciate.

4.3.1. Assume on-peak demand is

$$P = 16 - 0.08 \cdot Q$$

and off-peak demand is

$$P = 16 - 0.16 \cdot Q$$

If the existing capacity (Q) were 120 units, what would be the socially optimal prices during the on-peak hours and during the off-peak hours?

4.3.2. What would be the optimal capacity? What would be the prices at this optimal capacity? (This problem relies on Viscusi, Vernon, and Harrington, 2000, p. 394.)

EXERCISE 4.4. THE AVERCH–JOHNSON MODEL

Averch and Johnson (1962) assume that the utility maximizes profit subject to the rate-of-return constraint:

$$\text{Maximize } PR = P \cdot Q(L, K) - w \cdot L - r \cdot K \tag{4.9}$$

$$\text{Subject to } s = [P \cdot Q(L, K) - w \cdot L]/K$$

where PR is profit, P is the price of electricity, Q is the output (which depends on labor and capital through the production function), r is the utility's cost of capital, and s is the allowed rate of return. (Note: Averch and Johnson use only two inputs; similar results can be obtained with fuel as a third input.) Further, Averch and Johnson assume that $s > r$, i.e., that the allowed rate of return is greater than the cost of capital. This problem can be transformed using the Lagranian multiplier technique (a more rigorous solution relies on the Kuhn–Tucker conditions):

$$\begin{aligned} \text{Maximize } PR &= p \cdot Q(L, K) - w \cdot L - r \cdot K \\ &\quad + \lambda \cdot [s \cdot K + w \cdot L - p \cdot Q(L, K)] \end{aligned} \tag{4.10}$$

where λ is the Lagranian multiplier. If the constraint is binding, then $(s \cdot K + w \cdot L -$

$p \cdot Q) = 0$. To solve for the constrained profit-maximizing levels of capital (K) and labor (L):

1. Differentiate with respect to K, L, and λ
2. Set each derivative to zero
3. Solve simultaneously for K and L

These derivatives are

$$\partial PR/\partial K = P \cdot Q_K - r + \lambda \cdot (s - P \cdot Q_K) = 0$$

$$\partial PR/\partial L = P \cdot Q_L - w + \lambda \cdot (w - P \cdot Q_L) = 0$$

$$\partial PR/\partial \lambda = s \cdot K + w \cdot L - p \cdot Q = 0$$

where
$Q_K = \partial Q/\partial K$ is the marginal product of capital
$Q_L = \partial Q/\partial L$ is the marginal product of labor

Because $\partial PR/\partial \lambda = 0$, the ROR constraint is satisfied at maximum profit. Averch and Johnson show that the *Marginal Rate of Technical Substitution* (*MRTS*) of labor for capital (the rate at which capital can be reduced when one extra unit of labor is used so that output remains constant) is

$$MRTS = \{r - [(s - r) \cdot \lambda/(1 - \lambda)]\}/w = (r - \alpha)/w$$

If the allowed rate of return is equal to the cost of capital ($s - r = 0$), then $MRTS = r/w$, i.e., the profit-maximizing solution is levels of capital and labor such that the *MRST* is equal to the ratio of input prices. Further, Takayama (1993, pp. 212–219) shows $dK^*/ds < 0$, where K^* is the optimal level of capital, i.e., as s decreases (holding r fixed), the optimal level of capital increases.

If $s > r$, "[T]he firm adjusts to the constraint, then, by substituting capital for the cooperating factor and by expanding total output" (Averch and Johnson, 1962, p. 1056). Therefore, because the firm earns a "bonus" on capital equal to $(s - r)$, it has an incentive to use more capital than it should if it were behaving efficiently.

The size of the distortion depends on λ and $(s - r)$. The Lagranian multiplier can be interpreted as a measure of the sensitivity of profit to the ROR constraint. If $\lambda = 1$, then profit is simply $K \cdot (s - r)$. If $\lambda = 0$, there is no ROR constraint. So as regulation becomes more constraining, the closer λ is to 1.

Empirically, it is difficult to determine the value of λ. On the other hand, whereas it is possible to determine the allowed rate (s) of return from regulatory proceedings, it is more difficult to determine the underlying cost of capital (r). Further, even if these variables are known, it is difficult to determine the firm's efficient choice of technology given *uncertainty* in the price of inputs.

4.4.1. Because $0 < \lambda < 1$, plot against $(s - r)$ for $\lambda = 0.1, 0.2, \ldots, 0.9$, and 1.0, where $\alpha = [(s - r) \cdot \lambda/(1 - \lambda)]$. What happens as s approaches r?
4.4.2. If the regulator is wrong and sets $s < r$, how do the firm's incentives change?

EXERCISE 4.5. A REAL OPTIONS PRICING MODEL OF COGENERATION

4.5.1. Nonutility Generator Investment

The Public Utilities Regulatory Policy Act (PURPA) of 1978 opened the U.S. wholesale electricity market to unregulated independent power producers: cogenerators and small power producers, known as NUGs. How does a NUG evaluate the risk of investing in new capacity?

For example, consider a paper mill's decision to invest in cogeneration equipment (see Exercise 3.4.4). Under the usual investment criterion, the paper mill would invest if the NPV is positive, where NPV is equal to the discounted value of its net revenues (R) minus the cost of the investment (I) :

$$NPV = [\Sigma\, R/(1 + r)^t] - I$$

for $t = 0$ to T, where r is the firm's cost of capital.

Let the cost of the electricity generation equipment equal \$3000/kW for 50 MW of capacity, Q. The total investment, I, is \$150 million. Assume that (1) the equipment can be installed instantaneously, (2) there are no operating costs (e.g., waste steam is free), (3) the capacity factor, CF, is 0.70, and (4) the remaining life of the paper mill is 10 years, T.

The current price, P, to NUGs for electricity offered to the utility is \$80/MWh for 10 years at 8760 hours per year. (This is similar to prices paid under PURPA contracts.) With a discount rate, $r = 10\%$, the uniform series, present worth factor is 6.145. Should the paper mill invest at the current price? The NPV (from $t = 0$ to 10) equals

$$\begin{aligned} NPV &= [\Sigma\,(8760 \cdot Q \cdot CF \cdot P)/(1 + r)^t] - I \qquad (4.11)\\ &= [8760 \cdot 50 \cdot 0.70 \cdot 80 \cdot 6.145] - \$150{,}000{,}000\\ &= \$150{,}700{,}000 - \$150{,}000{,}000\\ &= \$700{,}000 \end{aligned}$$

The NPV is positive. So under traditional investment criteria, the firm should install the electricity generating equipment.

Note that if the project life is long, discounted net revenues can be approximated as a perpetuity (i.e., the firm receives the same revenues in each period forever; see Chapter 3). Then discounted net revenues are equal to (R/r) and $NPV = (R/r) - I$. If $R > r \cdot I$, then $NPV > 0$, and the firm should install the cogeneration equipment.

Also, the present worth factor for $r = 10\%$ is equivalent to a discount rate of 16.3% for a perpetuity.

Next, consider the situation in which the price this year is \$80/MWh. But next year, the price could rise to \$100 with probability $P = 0.6$ or drop to \$60 with probability $(1 - P) = 0.4$. These prices would then remain the same for 10 years. Should the firm invest or wait to see what happens to price? If the price increases to \$100 with a probability of 0.6, the *NPV* (for $t = 1$ to 11) would be

$$NPV = [\Sigma\,(8760 \cdot Q \cdot CF \cdot P)/(1 + r)^t] - I/(1 + r) \tag{4.12}$$

$$= [(8760 \cdot 50 \cdot 0.70 \cdot 100 \cdot 6.145) - \$150{,}000{,}000]/1.1$$

$$= [\$188{,}000{,}000 - \$150{,}000{,}000]/1.1$$

$$= \$34{,}545{,}000$$

[Note: here discounting is from $t = 1$ to 11, dividing by $(1 + r)$ shifts t from 0 to 10.] If the price falls to \$60 with probability 0.4, the expected *NPV* falls to about 13,440,000. In fact, if the price falls by \$0.40/MWh, *NPV* is negative. The firm would not invest (after waiting) if the price fell, so *NPV* is zero and the expected *NPV* is $0.6 \cdot \$34{,}545{,}000 + 0.4 \cdot 0 = \$21{,}000{,}000$. The price where *NPV* is just equal to zero is the *trigger price*. Here the trigger price is \$79.60/MWh.

By investing today, the firm gives up the option of investing in one year, when it might earn an expected *NPV* of \$21 million. The value of this option is the difference between the expected *NPV* of waiting and the *NPV* of investing today: \$21 million – \$0.7 million = \$20.3 million. When considering the option value of waiting, it is unlikely that the paper mill would invest, i.e., the firm would wait and see.

4.5.2. Nonutility Generator Investment Options

What *cost per kilowatt of capacity* would leave the paper mill owner indifferent between investing today and waiting? We can generalize this example by assuming that revenue (a function of price) changes over time so that percentage changes in revenue ($\Delta R/R$) follow a normal distribution with a mean of 0 and a standard deviation of σ. (Above, price had a mean of $84 = 0.6 \cdot 100 + 0.4 \cdot 60$ and a standard deviation of about 29.) We assume that R either increases or decreases by a fixed percentage in each period, (dR/R) is constant. If these changes are unrelated from period to period and if we can characterize the changes over short periods (dt) as $(dR/R) = \sigma \cdot dz$, [where dz follows a normal distribution with a mean of 0 and a standard deviation of $(dt)^{1/2}$], then R exhibits proportional *Brownian motion*. See Dixit and Pindyck (1994).

Under these assumptions, the firm will observe random projected revenues. Let H be a trigger value such that if $R > H$, the firm invests. For example, under traditional investment criteria, this trigger is approximately $r \cdot I$. This is because if $NPV = (R/r) - I > 0$, $R > r \cdot I$. The value of the project is

$$V_I = (R/r) - I \tag{4.13}$$

where V_I is the (net present) value of investing. If $R < H$, the firm waits. Dixit (1992, p. 113) shows that the value of waiting, V_W, can be represented as

$$V_W = B \cdot R^{\gamma} \tag{4.14}$$

where B is a positive constant and γ is a function of the discount rate (r) and the standard deviation of net revenues (σ).

What is B? When $R = H$, the firm is indifferent between waiting and investing. So, the value of investing, $R/r - I$, must be equal to the value of waiting, $B \cdot R^{\gamma}$. Equating these and substituting H for R,

$$(H/r) - I = B \cdot H^{\gamma} \tag{4.15a}$$

or

$$B = [(H/r) - I]/H^{\gamma} \tag{4.15b}$$

Therefore, B is a constant that equates the value of investing and the value of waiting: $R = H$.

Further, Dixit shows that γ is a solution to the differential equation that describes the Brownian motion of R through time. The solution involves a quadratic expression in γ:

$$\gamma^2 - \gamma = 2\, r/\sigma^2 \tag{4.16a}$$

where σ^2 is the variance of R. Solving for the positive root of this expression,

$$\gamma = 1/2 \cdot \{1 + [1 + (8\, r/\sigma^2)]^{1/2}\} \tag{4.16b}$$

If $\sigma^2 > 0$ (i.e., revenues are uncertain), then $\gamma > 1$.

What is the optimal trigger value H? The optimal value occurs where small changes in R are equal for investing and waiting at $R = H$. The derivative of Equation (4.13) with respect to R is $(1/r)$. Also, we differentiate Equation (4.14), the expression for the value of waiting, with respect to R. This is equal to $\gamma \cdot B \cdot R^{(\gamma-1)}$. Setting these equal and substituting $R = H$:

$$1/r = \gamma \cdot \mathrm{B} \cdot H^{(\gamma-1)} \tag{4.17}$$

Solving for H, we find

$$H = [\gamma/(\gamma - 1)] \cdot r \cdot I \tag{4.18}$$

The trigger level is $\gamma/(\gamma - 1)$ greater than $r \cdot I$, the trigger level of the traditional (but approximate) net present value criteria. The discount rate under the *real options* approach to investment is $[\gamma/(\gamma - 1)]$ greater than the unadjusted discount rate:

$$r_{\gamma} = r \cdot [\gamma/(\gamma - 1)] \tag{4.19}$$

Reconsider the case of the paper mill with continual changes in revenues. Let the discount rate equal the perpetuity-equivalent discount rate of 16.3% and the standard deviation of proportional changes in revenue equal 36% (= $29/$80). Under these assumptions, what are the values of γ, $\gamma/(\gamma - 1)$, r_{γ}, and H, where I = $3000/kW? (This exercise is based on Rothwell and Sowinski, 1999.)

CHAPTER 5

COMPETITIVE ELECTRICITY MARKETS

5.1. OVERVIEW

Throughout the world, restructuring and competition are being introduced into the electric power industry. The traditional organization based on vertically integrated utilities (generation, transmission plus distribution, and retail), publicly or privately owned, is moving toward a new structure of companies producing *unbundled* services: generation companies have been separated from transmission and distribution companies. In wholesale markets, several generation companies compete to sell their energy. Competition is also being introduced into retail sales, so customers can choose among different retail companies and retailers can compete for market share by offering competitive prices and new services. Transmission and distribution still are considered natural monopolies that must be regulated. This regulation ensures open, nondiscriminatory, and tariff-regulated *third party access* to the network for all market participants: generators, retailers, and customers.

The specific objectives of these restructuring processes vary from country to country, but several common characteristics driving the changes can be pointed out:

- The traditionally regulated industry led to high electricity prices.
- Cross-subsidization among customer categories created inefficiency.
- New efficient generation technologies with small unit sizes (e.g., 150–300 MW) and shorter construction periods became competitive, and they could be built by private companies.
- Nationally owned sectors needed investments that could not be funded by the state, so, new private initiatives were required.

This chapter introduces the main ideas needed to understand the mechanics of competitive electricity markets discussed in Chapters 6–9. In Section 5.2, short-run operational issues associated with wholesale energy transactions and the operation of the transmission system are presented. Competitive wholesale electricity markets have been traditionally organized with pool trading and centralized coordination through a market operator. Additionally, some restructuring models allow physical bilateral trades between market participants outside the pool. Whatever the wholesale market organization, the physical integrity of the bulk power system must be guaranteed to meet the required security conditions. Combined ownership and control of the transmission system is the cornerstone for ensuring system security under the new unbundled structure. See Stoft (2002, Part 3).

In Section 5.3, long-term planning and investment issues, related to system adequacy for meeting future demands, are presented. Planning and building new transmission lines requires new ways of regulating transmission companies and system operators. On the other hand, market incentives might not induce construction of enough new generation. So additional capacity payments to generators might be needed.

Finally, Section 5.4 describes issues associated with retail sales and distribution systems. Retail competition has been introduced through customer choice in selecting among different retailers. Distribution companies will continue to be regulated, but new regulation schemes based on incentives are being implemented to improve economic efficiency. For more on these issues, see Fox-Penner (1997) and Stoft (2002). Examples of how each one of these issues have been solved for different systems can be found in the following chapters. Throughout this chapter, references will be made to the cases of California, Norway, Spain, and Argentina. For more detail see Chapters 6–9.

5.2. WHOLESALE POWER MARKETS

The traditional power sector based on vertically integrated utilities (generation, transmission plus distribution, and retail) is moving toward a new structure of vertical disintegration. In a *wholesale power market,* all generators compete to sell to all distributors, or directly to customers and retailers if *retail competition* is allowed. Under this new unbundled structure of generation and transmission with competitive trading between wholesale market participants, the traditional pool operating functions have been more clearly defined and segregated into:

- *Market operation functions* related to energy trading, scheduling, and settlement of energy transactions in different time horizons (week, day, or hour ahead before physical transactions occur) and
- *System operation functions* related to operation and control of the bulk power system (1) to meet load and security needs with real-time dispatch to balance supply and demand, (2) to manage ancillary services to maintain system reliability, and (3) to manage transmission congestion, etc.

Regarding the organization of competitive wholesale electricity trading, many of the restructuring experiences (e.g., in the UK; Argentina; Chile; Victoria, Australia; and Alberta, Canada) have been based on *pool* trading with centralized coordination by a ***Market Operator*** (MO), who is usually the ***System Operator*** (SO). The MO has a monopoly in arranging all wholesale transactions that affect the physical flow of electricity in the bulk power system.

More recently, during the restructuring debate in California, and previously in Norway and Spain, a more flexible wholesale model was proposed and implemented. Besides pool-based trading, *physical bilateral trades* scheduling generation and demand outside the pool are allowed. The separation of market operation functions and system operation functions into two independent entities, the MO and the SO, has been used in the hybrid pool plus physical bilateral trading model. For instance, in California, the Power Exchange (PX) matched bilateral physical trades between generation and demand, and submitted these schedules to the ***Independent System Operator*** (ISO), who was responsible for system reliability and transmission congestion management. (Note: the PX was dissolved during the California electricity crisis of 2000–2001.) In Norway, where physical bilateral transactions are also allowed, Nord Pool is the MO and Statnett is the SO. In Spain, a MO and a SO were created as two independent entities. This separation allows the possibility of several entities conducting trading even under competition. This seems to be the general trend in most new restructuring experiences, including the recent reform of the UK wholesale market.

Competitive wholesale trading in a pool market and by bilateral trading is analyzed in Sections 5.2.1 through 5.2.3. Regulatory issues concerning *transmission ownership* and the creation of an *independent system operator* are discussed in Section 5.2.4. Finally, new ways of unbundling and procuring *ancillary services* (services necessary to maintain reliable operation) under wholesale competition, are presented in Section 5.2.5.

5.2.1. The Poolco Market

In a ***poolco*** or *"mandatory pooling"* market, all generators sell into a pool run by a MO (the MO can also be the SO.) Generators' physical sales of power and energy trades are all within the pool. Similarly, all power flows and purchase transactions are between buyers and the pool. If the market allows *wholesale competition* only, the buyers are distribution companies (distributors) who then resell the power at retail to customers. If the market allows *retail competition*, buyers could be individual customers, suppliers, retailers, etc. See Stoft (2002, p. 224).

The MO holds an auction in which each generator bids different *prices* for different *quantities* of power (from specific plants or as a portfolio) for the trading period; for example, for each hour of the following day. Based on the bids and considering demand, the MO uses a matching process to set the market price (generally one price per hour) and the generation quantities.

One of the most popular market clearing systems is the *simple matching algorithm*, in which the MO first chooses the quantity with the lowest bid price and uses as much of it as possible. Next, the MO does the same for the quantity with the next lowest price, and so on until demand is covered (see Exercise 5.1). (Under price competition, costs are private information and prices are public information.)

Buyers can also bid quantities and prices. The MO ranks them in a demand curve by decreasing price. The "winners" of the auction (i.e., all plants and all customers that will be scheduled, represented by the intersection of the supply and demand curves) are announced. Usually, the price of the last generator scheduled becomes the system-wide pool price. This price is known as the ***market-clearing price*** (MCP): the price that *all* sellers will receive and that *all* buyers will pay. This market mechanism is the basis of poolco competitive generation markets. Figure 5.1 shows market agents and transactions in a poolco market. (Market power issues in competitive generation markets are discussed in Section 5.3.1.2.)

In wholesale power markets, as the bidding sessions get closer to real-time dispatch the SO takes more responsibility. For instance, the SO (1) takes into account the trades that were concluded by the MO (sometimes, the SO and MO are the same) and (2) checks (and makes adjustments) to ensure system security. For example, the SO can require some units running at lower power to maintain spinning and standby generation reserves or the SO can modify the dispatch order to avoid congestion in the transmission system, etc. In some cases, transmission constraints can lead to different area or zonal market clearing prices, as in California (see Section 5.3.2). Examples of wholesale poolco markets have been

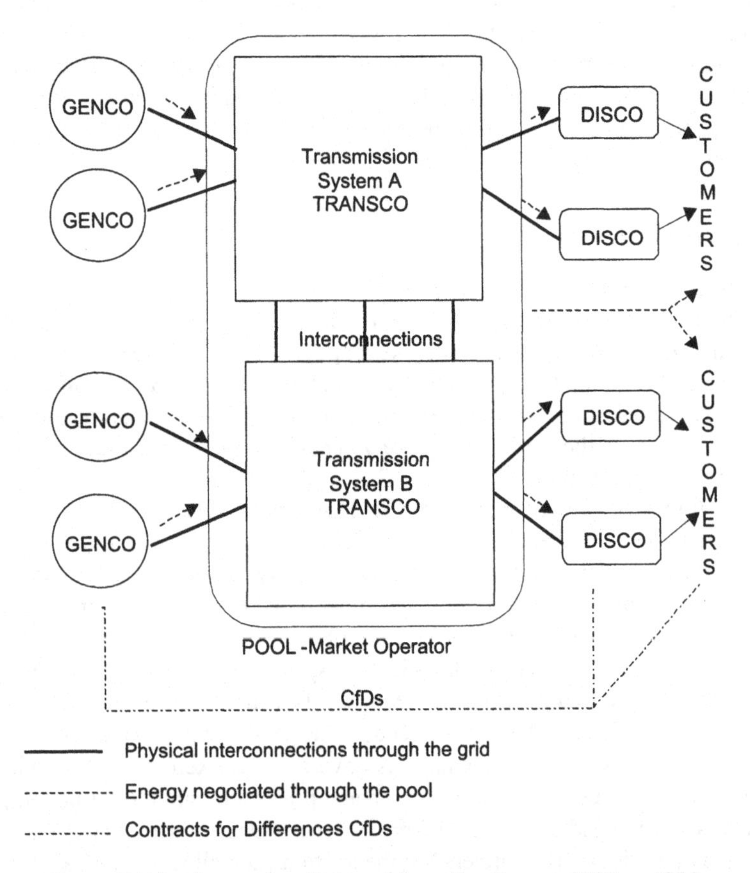

Figure 5.1. Poolco market. (Adapted from Fox-Penner, 1997, p. 185.)

1. The daily energy market in the UK before the NETA reform
2. The day-ahead and hour-ahead markets run by the PX in California
3. The day-ahead and the intradaily energy markets run by the MO in Spain. (Note that in Spain and California, the possibility of physical bilateral trades is also allowed.)
4. The economic dispatch run each hour by the SO in Argentina

See case studies in Chapters 6–9 and Stoft (2002, Part 3).

5.2.2. Contracts for Differences

Pool prices signal (1) to generators whether more or less power is needed and (2) to buyers the cost of marginal generation. In general, low short-run elasticity of electricity demand and a steep generation supply curve produces prices that could rise and fall dramatically with daily and seasonal load variations (as they did in California during 2000 and 2001). Deregulation puts the market participants in a more risky position than under regulation. This *price volatility* leads buyers and sellers to make contracts that specify prices and supplies in advance with more certainty.

The most popular contract to reduce market uncertainty in a poolco is the ***Contract for Differences (CfD).*** See Stoft (2002, p. 211). CfDs are two-way wholesale contracts appropriate for buyers and sellers who trade directly in the pool. They guarantee a *price* and *quantity* for power between a buyer and seller. The "seller" agrees to sell to the "buyer" a schedule of power at specified prices over the term of the contract. This contract guarantees that the "buyer" gets the contracted quantity and price. However, under the poolco approach, unlike in physical bilateral trades, the way this occurs is not with an actual power sale between the "seller" and the "buyer." Instead, the "buyer" continues to buy all power through the pool, including the amount contracted for in the CfD. If the "seller" is a generator, it can be scheduled (or not) by the pool according the auction results. In this sense, the CfD is a *financial instrument.* It does not involve a *physical power transaction* between a "buyer" and a "seller" with specific scheduling of the seller's generation. All the "seller" does in the CfD is

1. Compute the difference between the price it has guaranteed to the "buyer" and the price the "buyer" paid to the pool
2. Pay the "buyer" the difference (if positive) or collect it from the "buyer" (if negative)

Exercise 5.4 explores the mechanics of a CfD. Other financial instruments, such as ***futures*** and ***forwards*** (see Chapter 7, this volume; Kaminski, 1997; and Hunt and Shuttleworth, 1996), and corresponding markets, have been proposed or are under development in electricity markets.

5.2.3. Physical Bilateral Trading

In some wholesale electricity markets (such as in Norway, California, and Spain) in addition to voluntary pool trading with the possibility of financial ***bilateral contracts***, physical bilateral trades (i.e., purchases between contracting parties) are also allowed. Buyers and sellers individually contract with each other for power quantities at negotiated prices, terms, and conditions. Generators can directly contract with buyers and purchase transmission services from the transmission owner or system operator. Supplier companies (marketers) can make agreements to aggregate, bundle, or sell power from generators to the transmission system or to distributors directly. If retail competition is allowed, generators and marketers could also make contracts with customers directly. All transactions must be announced to the SO, which analyzes all the trades in each period and determines, without discrimination and with some clear rules of prioritization, which ones are infeasible due to grid security constraints. The SO need not know the prices. Figure 5.2 shows market agents and transactions involved in bilateral trading.

In a pure bilateral model based only on physical bilateral trades, neither homogeneity of the transactions nor a single market-clearing price is ensured; see Exercise 5.2. Buyers and sellers must shop and discover prices through advertising, market information services, and comparison shopping. Centralized trading, as in a poolco market, can lead to significant savings by avoiding the need to discover prices, because the pool market price is publicly known. In theory, centralized trading in a pool can be more efficient because most market participants are scheduled and tracking agreements between pairs of traders is not necessary.

However, practical implementation issues in favor of physical contracts are

1. Centralized coordination creates general management costs that should be shared by all market participants (trading outside the pool avoids these costs).
2. Bilateral supporters do not trust the market clearing process to function adequately and they want to ensure their generation through physical contracts.

The last trend in wholesale market design allows simultaneous pool and physical bilateral trades.

On the other hand, physical bilateral trades and MO pool schedules have associated energy imbalances, i.e., mismatches between the contracted or scheduled quantities and the actual demand read from the meter. Because power consumed and supplied must be equal at every moment, there must be a mechanism to supply the balancing amount of power instantly. The source of power for settling energy imbalances can be a spot market for imbalances managed by the SO and created by pooling all available surplus generation. This market functions with the same rules as in the poolco energy market, where bids are for incremental quantities.

In Norway, there are wholesale physical bilateral contracts and there is a pool market for settling real-time imbalances. In California, there is a pool-based, real-time energy market for energy imbalances. In Spain, physical bilateral contracts are allowed together with pool energy transactions; these contracts must be announced

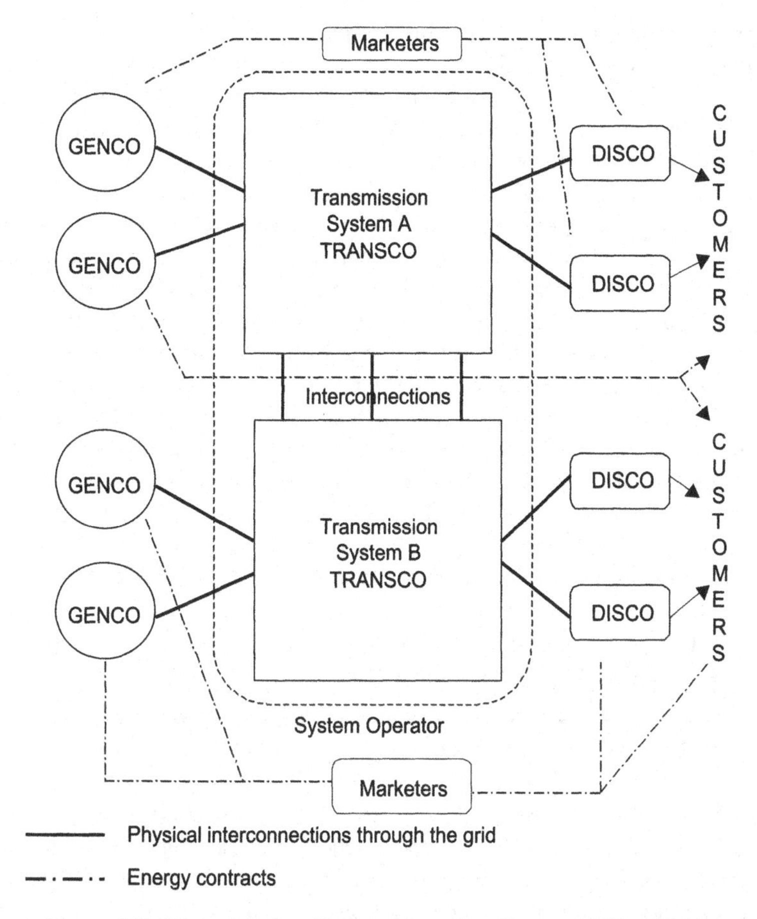

Figure 5.2. Bilateral trades. (Adapted from Fox-Penner, 1997, p. 186.)

to the market and system operators. In the UK, before the *New Electricity Trading Arrangements* (NETA) reform (Green, 1999) physical bilateral contracts outside the pool were not allowed. In Argentina, physical contracts have not been allowed.

5.2.4. Transmission Ownership and System Operation

An important issue to ensure effective wholesale electricity competition concerns the division of responsibilities into owning, operating, and regulating the transmission system. A cornerstone of restructuring is the separation of generation from transmission. Under this segregated structure, transmission companies and system operators continue to be regulated. All *transmission system owners* (TRANSCOs) must provide comparable and nondiscriminatory service to independent generators. Transmission has the cost structure of a natural monopoly (see Chapter 2).

The availability of transmission installations and the location of new transmission investments significantly affect trading opportunities. On the other hand, TRANSCOs must receive revenues that adequately compensate them for operation and investment costs. Transmission costs should be allocated to provide accurate locational price signals to generators. Also, the system operation function must continue to be regulated. Services related to system operation also have monopolist characteristics that affect the transparency and efficiency of the energy market.

Two approaches have been adopted regarding transmission ownership and system operation. The first approach is to separate transmission ownership from system operation. In some restructuring experiences, competition is being introduced into a vertically integrated industry, where transmission owners typically own generation and there are multiple owners of the interconnected transmission grid. This is the basis for the creation of an ISO. With an ISO, one or more TRANSCOs own the transmission system. But a single entity, the ISO (ideally independent of all market participants), operates the transmission system. This separation works as follows:

- The ISO dispatches generators according to the pool auction or bilateral contracts.
- The ISO contracts with the TRANSCO to enable it to use, charge for, and entirely control all of the TRANSCO's assets.
- TRANSCOs do not control dispatch or transmission, but receive regulated rates for the use of their assets.

The ISO needs such control to prevent transmission owners (who might also own generation capacity) from, for instance, scheduling transmission maintenance to raise generation prices. In California, an ISO was created but transmission assets remained the property of the three main investor-owned utilities.

As an example of ISO functions and responsibilities, the U.S. Federal Energy Regulatory Commission (FERC) requires the following before approving of a new ISO:

- A fair, nondiscriminatory governance structure
- No financial interests by any power market participant
- A single, open-access tariff for the entire ISO area
- Responsibility for system security
- System control for pool or bilateral dispatches
- Identification and resolution of transmission constraints
- Incentives to act efficiently
- Transmission and ancillary service pricing that promotes efficiency
- Transmission availability in real time on electronic bulletin boards
- Coordination with adjacent control areas
- A dispute resolution procedure

In other restructuring experiences, although transmission owners do not own generation, there is a separate ISO to allow the possibility of several transmission owners. This is the case of Argentina, where TRANSENER owns most of the transmission network and CAMMESA (Compañía Administradora del Mercado Eléctrico Mayorista SA) operates as the SO.

With the second approach, transmission ownership and system operation remain together. Usually, in these cases, there is a *national* transmission company that acts as the SO. Defenders of this scheme argue that it is difficult to implement strong performance incentives on an entity (the ISO) with no assets and little accountability. Under this scheme, the transmission system's part of the company continues to be regulated through, for instance, *performance-based ratemaking* (see Chapter 4), allowing it to make a profit. This level of profitability could be used to reward good performance. This unified scheme with a single entity as transmission owner and SO has been adopted in the UK, where the National Grid Company (NGC) performs both activities. In Norway, Statnett is the national grid owner and SO (see Chapter 7). In Spain, REE acts as the national transmission utility and SO (see Chapter 8).

5.2.5. Ancillary Services

The services necessary to support a reliable interconnected transmission system are *interconnected operation services* or ***ancillary services***. See Stoft (2002, pp. 232–242). Across the different restructuring experiences, these services are not uniquely defined. Their unbundling and procurement mechanisms are greatly influenced by the overall organization of a specific electricity market. For instance, in the US, FERC defines the following ancillary services required in open-access transmission systems:

- *Scheduling, system control, and dispatch*: management of system operation
- *Regulation and frequency response*: generators instantly increase or decrease output to match load fluctuations
- *Operating reserve (spinning reserves and supplementary reserves)*: generators are kept warm or on standby, ready to take over following generator or transmission line failure
- *Energy imbalance*: residual energy that must be provided to settle the imbalance between energy contracted and actual energy delivered
- *Voltage control and reactive power support*: control equipment such as transformers, etc., needed to maintain network voltages within security limits
- *Loss compensation* for the difference between total energy generated and total energy delivered to all customers, i.e., network losses

Regarding payment for these services, payments depend on the type of energy transactions, *bilateral* or *pool*, that have been scheduled. Bilateral transactions are charged according to their impact on the costs of providing the associated services. In a poolco market, the cost of providing these services can be added to the hourly spot price, whereby all customers pay an additional bundled charge with the hourly energy charge, or they can be charged separately. (Energy imbalances are typically charged separately.) The determination of separate charges depends on (1) whether the participant who caused the cost (or the saving) can be identified and (2) whether the contribution can be valued.

Regarding the provision of ancillary services, first, some services (for instance, scheduling and dispatch) can be only provided by the SO. Second, other services (e.g., operational reserves, reactive power support, etc.) can be provided by others, for instance, generators or customers with controllable loads. In the first case, the monopoly service provided by the SO and its price must be regulated. In the second case, the SO is also responsible for system security, so it obtains the required amount of each service from the various suppliers.

System operators use various procurement schemes. Some services are made compulsory. Others can be procured through short-term bids (e.g., operating reserves), or through long-term contracts, usually for one or two years (e.g., reactive power support). Long-term contracts are recommended for services that can be provided by only a few suppliers. For example, for reactive power support, only generators or sources electrically close to the voltage problem area are able to provide local support.

In the UK, the procurement procedure for some services is based on long-term contracts between the NGC and the generators. In California (Chapter 6) and Spain (Chapter 8), the ISO conducts day-ahead bidding auctions to buy regulation and operational reserves from generators. In Norway (Chapter 7), generators have an obligation to provide ancillary services through contracts with Statnett; they receive payment only if quantities provided are beyond contracted ones. In Argentina (Chapter 9), there is an obligation for each generator that sells energy to provide a specific level of ancillary services. However, CAMMESA uses a competitive weekly bidding auction to purchase "cold" operating reserves from generators. These ancillary services market experiences are so recent that it is too early to conclude how best to organize them.

5.3. MARKET PERFORMANCE AND INVESTMENT

Assuming that the bulk power system continues to operate on a daily basis with adequate efficiency and reliability, at some time new generation capacity and transmission lines will be needed. In this section, we examine conceptual and regulatory issues associated with the long-term *adequacy* of generation and transmission. *Adequacy* is the ability of the power system to meet demand, taking into account scheduled and reasonably foreseen unscheduled outages of generators and transmission installations. Deregulation of prices must create market incentives sufficient to stimulate generation expansion in the long run. However, oversight of the market to ensure effective competition among generators is still a primary regulatory task.

Also, the expansion of the transmission system to provide long-run economically efficient and reliable service in the presence of competitive generators must be ensured by adequate transmission regulation.

5.3.1. Generation Expansion and Monitoring Generation Competition

In theory, the output of a competitive generation market is equal to the output of a regulated system with a central planner that minimizes investment plus operating costs to meet demand (Green, 2000). However, in a market the uncertainty of recovering generation investment plus operating costs is higher than under regulation, under which generation revenues were guaranteed. The *price volatility* of power markets is the rationale for both sellers and buyers to make *long-term contracts* to hedge against uncertainty. These contracts usually do not last more than a few years and they are rarely the basis on which financiers award construction loans for generation plants. More frequently, financiers back plants because the borrower (typically an established utility that can assume risk) can provide collateral. However, the expectation of future cash flows is currently the main market force driving generation expansion in competitive markets.

In these markets, generation capacity should be built so that the *long-run marginal cost of electricity equals the average revenue* from market sales. Building capacity above the optimal level will depress average spot prices and depress average revenues, and some generators will leave the system. Less capacity will increase average revenues, attracting new capacity to the system. However, if demand is not allowed to set the market price (e.g., under load curtailment situations), then market prices, average revenues, and investment in capacity could be less than the efficient level (see Exercise 5.3). Demand flexibility is especially important under capacity constraints, as the experience in California has shown.

5.3.1.1. Generation Revenues and Capacity Payments. Under a regulated monopoly, compensating generators based on the load factor of each plant is equivalent to paying each generator its technology-specific capacity and energy costs. This can be observed on the left-hand side of Figure 5.3, where three different generation technologies are dispatched to cover a load duration curve. The generator with the highest fixed costs and the lowest variable cost, GN3 (a base-load generator), receives its fixed (F3) and variable costs (C3) for producing energy, $F3 + C3 \cdot (T1 + T2 + T3)$, where $(T1 + T2 + T3)$ is its total generating time. Consumers with the highest load factors pay this amount. The generator with the lowest fixed cost (F1) and the highest variable cost (C1), GN1 (a peaking generator), receives its fixed and variable costs during the period it is producing energy, $F1 + C1 \cdot T1$, where T1 is its total generating time. This amount is paid by the customers with the lowest load factors. In this way, generation revenues cover the fixed and variable costs of each technology. See Oren (2000).

However, in a competitive generation market, in each period generators receive the variable cost of the most expensive generator dispatched in that period. On the right-hand side of Figure 5.3, we see that in hours with higher load factors, the mar-

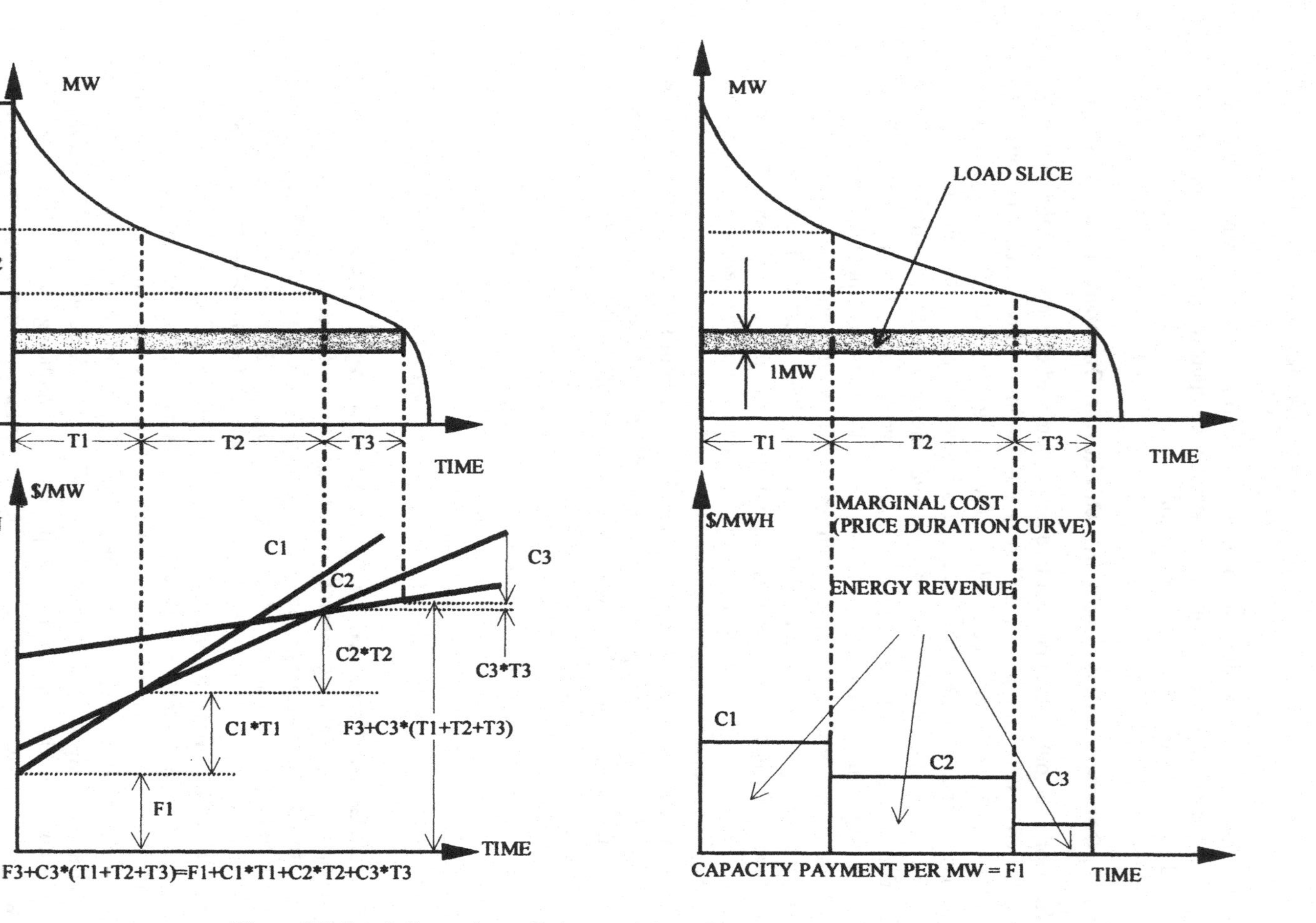

Figure 5.3. Load slice and marginal cost pricing. (Source: Oren, 2000.)

ginal energy cost (market price) is equal to the lowest variable generation cost (C3). In hours with lower load factors, the marginal energy cost is equal to the highest variable generation cost (C1). Therefore, load slices with the highest load factors will pay (C1 · T1 + C2 · T2 + C3 · T3), and load slices with the lowest load factors will pay (C1 · T1).

We can see that the sum of the marginal cost times the duration during which it applies ($\Sigma\, C_j \cdot T_j$) is equal to only the variable costs of producing electricity. Thus, if generators are paid a uniform marginal cost for their energy, they end up with a shortfall in the amount of F1, the fixed cost of the peaking technology. This is the reason that some systems, as in Chile, have introduced an explicit ***capacity payment*** to generators equal to the fixed cost of the peaking technology.

Under optimal generation planning, the marginal cost of incremental generation capacity equals the marginal cost of unserved load. One approach to capacity payments attempts to equate this payment to the marginal cost of the unserved load. This can be approximated by (1) the marginal value of unserved load (or *value of loss of load*, VOLL) times (2) the *loss of load probability* (LOLP) or fraction of time that load must be curtailed due to insufficient capacity. England and Wales followed this approach before the NETA reform to supplement generator marginal energy revenues.

In theory, the need for a capacity payment for generation investment recovery could be eliminated (1) by allowing demand-side bidding in the market or (2) by allowing demand to set the market price with insufficient capacity. Demand bidding is equivalent to setting the spot price at VOLL during curtailment periods in which the true value of lost load is revealed. (See Exercise 5.3 for more details on generation cost recovery and long-term market equilibrium in competitive generation markets.)

In practice, even with demand bidding, it might not be politically feasible to allow electricity prices to fully reflect scarcity rents. Consequently, energy prices are often suppressed through regulatory intervention with price caps or by market design. Therefore, capacity payments are introduced as remedial actions to offset revenue deficiency and to encourage capacity investment.

Different approaches to capacity payments have been followed. In some experiences, as in Chile, Argentina, Colombia, Bolivia, Peru, and Spain, explicit capacity payments have been introduced as additional remuneration to the pool energy payments received by the generators. An administrative price (US$/kW) has been set to remunerate the *firm* capacity that each generator can supply in periods with higher probabilities of energy shortages. To allocate these payments, in energy-constrained systems based on important hydroelectric production, dry hydrological conditions are considered. In capacity-constrained systems, primarily based on thermal units, peak hours and forced outages represent the most likely conditions of shortage. In the UK before the NETA reform, the pool energy price was complemented with an additional payment equal to the marginal VOLL times the LOLP.

In other experiences, such as in Australia, New Zealand, Norway, and California, there are no explicit capacity payments. It is assumed that market prices and demand-side bidding through bilateral contracts will ensure enough generation investment *in the long run*. Other approaches, as in the PJM (Pennsylvania–New Jersey–Maryland Power Pool) and in the New York Power Pool, use market-based mechanisms to contract for long-term supply. See Stoft (2002). In Scandinavia, markets for peak power are under development. For more detail about capacity payments, see Oren (2000) and Vazquez, Rivier, and Pérez-Arriaga (2002).

5.3.1.2. Market Power Issues. Given the market structure of electricity, market monitoring and surveillance in generation is centered on *market power* issues, defined as the ability to profitably raise price by "a small but significant and nontransitory amount" (DOJ/FTC, 1992). Insufficient or ineffective competition among generators will yield profits and prices that are higher than under effective competition, which would undo the benefits of restructuring. This effect appears when a few firms can dominate generation markets. Restructuring experiences suggest that *generation concentration* is an important concern.

For instance, in California, regulators required the largest utilities to sell some of their generating plants to ensure that the emerging power market was competitive. However, market power appears to have been exercised in California markets to raise prices by a significant amount for several months in 2000 and 2001 (see Chapter 6). Therefore, oversight of the market by regulators and adequate antitrust enforcement are necessary to counteract market power problems. The economic foundations of market power in electricity markets are studied in Green (2000). The practical issues associated with market power can be found in Borenstein, Bushnell, Kahn, and Stoft (1995).

5.3.2. Nodal and Zonal Transmission Pricing

Three approaches to transmission pricing have been used: *uniform pricing, nodal pricing,* and *zonal pricing*. The UK and Spain use ***uniform pricing,*** i.e., there are no transmission price differences within a market. This pricing approach is appropriate for electric systems in which the transmission network is well connected and there are no structural congestion problems.

In other experiences, such as in Argentina and Chile, where there are long transmission systems and great distances between supply and demand areas, ***nodal pricing*** has been applied. For example, Argentina calculates nodal prices as (1) *nodal factors* that affect the energy spot price in each node and (2) *adaptation factors* that affect the generator's capacity payments in their node. The PJM (US) also uses nodal prices. See Stoft (2002, pp. 424–430).

The value of transmission between two nodes (or areas of the system) is given by the difference between energy prices of the two nodes. The price difference appears (ignoring transmission losses) because of flow limit constraints imposed by transmission capacity. Thus, the cost of meeting demand in various geographic areas differs. Nodal prices can be calculated at each transmission node as the marginal cost of meeting an increase in demand at that node. Nodal prices consider (1) the marginal cost of transmission losses and (2) if transmission congestion exists, the cost of extra generation (more expensive than the system marginal generator) that must supply the demand increment. For this reason, nodal prices depend on the characteristics of the transmission network. Nodal prices send market participants locational price signals regarding transmission losses and transmission congestion. Under nodal pricing, every major transmission node (generator, transmission line junction, or substation) has a spot energy price. See Schweppe, Caramanis, Tabors, and Bohn (1988) for more details on spot energy prices.

Under deregulation, the following drawbacks have been pointed out regarding the use of earlier optimization tools for calculating nodal prices (for more details, see Exercise 5.6):

1. Nodal prices are extremely sensitive to data, particularly to line flow limits that depend on complicated, and sometimes subjective, technical considerations.
2. Congestion in a line produces nodal price differences in other noncongested lines.
3. Power can sometimes flow from higher- to lower-price nodes due to nontransparent mechanisms.

An alternative to nodal pricing is ***zonal pricing***. Under zonal pricing, nodes are gathered into zones that are bounded by potential constraint interfaces. Each zone has a spot energy price. The purposes of zonal pricing are (1) to encourage generators to locate within the boundaries of the high-priced zones and (2) to focus attention on relieving the flow constraints in the congested interfaces between zones. In Norway (see Chapter 7), there are (1) *marginal loss percentages* for each transmission node, (2) *transmission capacity charges* for different system areas, and (3) *zonal prices* for congestion management on major transmission lines. California also uses zonal pricing (see Chapter 6). However, defining zone boundaries (because of their dependence on operating conditions) can be a difficult task. Further, boundaries must be updated from time to time. See Stoft (1998 and 2002, Part 5) for more details on nodal and zonal pricing.

5.3.3. Transmission Planning and Investment

Under the new structure of generation and transmission segregation, transmission companies (TRANSCOs), must plan, budget, finance, and construct the transmission network to

1. Accommodate new generators
2. Provide for robust long-term competition
3. Maintain reliability
4. Consider investments in generation versus transmission

Under ideal competitive conditions, each node of the transmission system would have a nodal price. If price differentials persist, suggesting that generating capacity is consistently cheaper in one place than another (e.g., due to transmission congestion) transmission expansion would be merited. The final test for transmission expansion is whether operating cost savings resulting from expanding the network and relieving congestion are greater than the investment cost (see Exercise 5.5). Even where there are no price differentials between nodes, providing for higher reliability requirements will drive transmission expansion. Moreover, common capital costs in transmission are likely to lead to continued central coordination and regulation.

Regulation of transmission companies must allow for adequate revenues to compensate them for operation, maintenance, and investment costs. These revenues are collected from transmission users as transmission charges. Transmission charges have been instituted as access charges, connection charges, use-of-system charges, etc. They have been decomposed into fixed charges and/or variable charges (U.S.$/kW or U.S.$/kWh). They are paid only by consumers or by consumers and generators (see the following chapters for discussions of transmission charges in California, Norway, Spain, and Argentina).

Also, in some restructuring experiences with nodal or zonal pricing, ***congestion charges*** are collected by the SO and paid to the transmission owner. In California, for instance, a *usage charge* is calculated for congested transmission lines or interfaces and collected by the ISO. The usage charge is determined as the marginal value of the congested interface: if one MW of additional transfer capacity would be available at the congested interface, how much would the system marginal cost decrease. Some approaches have proposed calculating the usage charge as the price difference between the two nodes or zones linked by the congested interface. In systems with parallel paths (nonradial systems) the usage charges are not always equal to the nodal price differences between the congested areas (see Exercise 5.6). The total system congestion charges can be calculated as (1) the sum of the usage charges times the congested line flows or (2) as the sum of the price differences times line flows. However, the allocation of these congestion charges should be calculated by considering the usage charge of each interface as the marginal value of the congested interface.

The following two decisions are independent: (1) whether nodal or zonal pricing should be adopted and (2) whether congestion rents should be kept by the transmission company as part of its total revenues. From a practical perspective, total congestion charges should not influence the total remuneration of the company. Otherwise, the transmission company will have incentives to "create" congestion: if the grid were not adequately developed, price differentials would be higher and congestion rents larger.

On the other hand, financial network contracts (Hogan, 1992) or ***Firm Transmission Rights*** (FTR) have been proposed (1) to hedge market participants against congestion charge volatility and (2) to provide incentives to invest in transmission capacity. A FTR is a contractual right that entitles the holder to receive a portion of the usage charges collected by the SO when congestion exists, whether or not the holder of the FTR transmits through that interface. See Stoft (2002, pp. 431–441).

For instance, assume that a remote generator with a low marginal production cost builds a line to sell energy to a more expensive load area, receiving all the

FTRs on the line. During an initial operation period, the line is not congested and congestion charges are zero. The holder benefits, however, from the high price of the electricity in the load area. This justifies investment in the line. After this initial period, a new, cheaper generator is installed close to the remote generator. The new generator manages to sell energy to customers using the same line, thereby congesting the line. In this situation, the generator who owns the FTRs collects congestion charges. In other words, FTRs ensure that those who invested in the line receive compensation even when they do not use it.

Additionally, FTRs allow market participants to hedge the risk of transmission price fluctuations. For example, a generator and a customer connected through an interface enter into a fixed-price supply contract. For that, they buy the FTRs of the interface that match the MW size of the transaction. This is the transmission price, whatever the actual congestion charges. The expected congestion charges would equal the FTR payment. If actual congestion charges are higher than the FTR payment, then the FTR holder makes a profit, and the FTR seller would suffer a loss of profit opportunity (assuming that the seller is not the SO, which is neutral to collect congestion rents), and vice versa. So, FTRs can be traded in secondary markets at the expected net present value of congestion charges for the contract duration. For more information on transmission management and congestion in electricity markets see Christie, Wollenberg, and Wangensteen (2000).

Finally, we address the regulatory issue of transmission ownership separation from system operation. For instance, when the national transmission company and the SO form a single, regulated, independent entity, transmission planning is improved with the information gained from system operation. In the UK, Spain, and Norway, there is a single SO that owns the national transmission grid and carries out transmission planning and investment. In Argentina, the SO is in charge of transmission planning, whereas transmission companies build new installations for users that benefit from the additions. In the United States, transmission grids are owned by formerly vertically integrated utilities; the role of new ISOs regarding transmission planning has not been clearly established. On the other hand, electric restructuring has intensified the interest in, and the need for, region-wide transmission planning and regulation in the United States.

5.4. CUSTOMER CHOICE AND DISTRIBUTION REGULATION

Many restructuring experiences coincide with the introduction of ***customer choice*** or ***retail choice,*** i.e., allowing consumers to choose their electricity supplier. Retail choice often means that the consumer is allowed to buy energy directly in the wholesale market (if it is a large user) or, more likely, from an electricity provider in the retail market. Retail choice does not mean that the consumer ceases to be connected to the grid, or that the consumer gets connected to a different company. The consumer will continue to pay a retail access tariff for the grid use. This retail access tariff is (1) regulated, (2) can have the same structure as the full tariff (though its level will be lower since it will no longer include the cost of electricity), and (3) can include any number of costs that the regulatory authorities decide to charge to electricity consumers.

Industrial customers promote customer choice to lower costs because they can buy energy at prices below former regulated tariffs. Independent generators see it as a way to enter the market by directly contracting with potential buyers. Customer choice brings new competitive market participants, *retailers* or *Energy Service Providers* (ESPs), that offer a variety of retail services with the sale of electricity.

Implementing customer choice differs from country to country. There are two basic alternatives:

1. Progressive implementation that starts by permitting only large customers to choose their supplier, followed by a gradual implementation for smaller customers over several years, ending with residential customer choice
2. Immediate implementation for all customer classes

In Norway and California, retail choice was instituted immediately upon deregulation (but was discontinued in California during the energy crisis of 2000–2001). Most of the deregulation experiences have adopted a progressive implementation over several years. This means the coexistence of a competitive sector to supply deregulated customers and a regulated sector (usually regulated distribution companies) to supply still captive customers. A gradual implementation allows utilities to make progressive adjustments to facilitate the transition from regulated to market revenues. Immediate retail choice allows customers to get the benefits of competition sooner and avoids cross-subsidizing captive and deregulated customers.

Under retail competition, some aspects of retail continue to be subject to regulation. Reliability and service quality will continue to be the responsibility of regulated distribution companies, because distribution companies can best ensure quality. Programs in energy efficiency and in renewable energies can continue to be regulated. Costs of these programs can be recovered through a retail access tariff. The regulator should carefully determine restructuring costs that residential customers pay in comparison to other types of customers through the design of a retail access tariff. *Distribution companies* (DISCOs) will remain regulated with or without retail choice. DISCOs will continue to be responsible for operating and developing distribution networks and serving regulated customers.

Metering and billing services are also being unbundled and exposed to competition. DISCOs are losing the monopoly of carrying out these services and new agents (such as retailers) are entering this area. Finally, the integration with other retail services, such as gas and water supply, telephone, TV cable, etc., are ways of making electricity retail businesses profitable.

5.4.1. Customer Choice and Retail Competition

Customer choice is a form of retail competition. A prerequisite for successful customer choice is a well-functioning wholesale market and the associated institutions, described in the previous sections. The agreement between a *retail customer* and the *retail company* is unregulated and can be structured in many ways. For instance, the price of service can vary by time and season, or other financial and quality of service conditions can apply. Retailers can contract with generators, or they can choose to buy energy directly from the wholesale market or from other supplier companies, such as *aggregators*. The retailer pays the wholesale market (or its own supplier company) for the energy consumed by its retail customers. In addition, retailers can arrange for the provision of network services, or the customer can arrange for those services directly. There are examples where customers sign separate contracts (1) for retail energy, (2) for network services, and (3) for the provision of metering services.

Therefore, depending on the form of restructuring, retailers can

1. Arrange for supply to meet the customers' loads
2. Arrange transmission and distribution network services and comply with established network procedures
3. Arrange contracts with ancillary services providers to meet current standards for reliability, backup supply, voltage regulation, operating reserves, etc.
4. Match supply with customers' loads, and account for losses and imbalances
5. Schedule sources and demands with the SO

On the other hand, in the UK, for example, retailers do not perform any of these functions. The pooling arrangement ensures the performance of items (1), (3), (4), and (5), and the customer makes the necessary network arrangements (item 2).

Regarding energy imbalances, retail contracts (1) tend to be "fixed-price, variable-quantity," (2) do not involve a demand forecast, and (3) usually do not participate in energy imbalance settlements. Only those agents (retail customers or retailers) that submit demand forecasts or demand bids can incur imbalances and, therefore participate in these settlements.

Regarding *billing*, there are many options available. For instance, retailers or ESPs could offer three billing options:

1. Consolidated *Utility Distribution Company* (UDC) billing, in which both the ESP's energy charge and the UDC's network service charge are billed by the UDC
2. Separate billing, in which the ESP and the UDC send separate bills for their respective services
3. Consolidated ESP billing, in which the ESP bills both ESP and UDC charges

Retailers do not have direct control of network reliability. As mentioned above, network reliability (e.g., associated with interruptions of supply due to network outages) is the responsibility of transmission and distribution companies. The standard reliability level must be regulated. In a few cases, retailers must negotiate reliability levels with the distribution company, either because the customer wants a different standard of reliability, or because the customer can negotiate directly with the distribution company.

Finally, customer choice and retail competition have little effect on the economic efficiency of the distribution system. Economic efficiency of distribution is related to the efficient segregation and allocation of its costs through adequate design of retail access tariffs (see Section 5.4.3) and to the regulation of DISCO revenues (see Section 5.4.4).

5.4.2. Real-Time Prices and Retail Services

Time-sensitive electricity rates, based on estimates of cost differentials, have been used under traditional regulation for all types of customers, even residential customers with meters that discriminate between daytime and nighttime consumption.

Under retail competition and customer choice, some approaches employ remote metering systems with ***real-time prices*** based on wholesale-market hourly prices, particularly for industrial and large commercial customers. Users who can manage their loads as a function of electricity prices can benefit from real-time pricing. Market prices for each pricing period of the next day can be communicated to all market participants. Any customer (even a small one) with a computer and Internet access can obtain this information from the MO's web-site.

Alternatively, ***load profiling*** can be used with small residential customers without a remote hourly metering system. In Norway, for instance, customers with annual consumption below 500 MWh are invoiced based on estimated load profiles (see Chapter 7; also see Detroit Edison, 1998). But the use of this technique will be reduced with the expansion of two-way communication systems and lower prices of hourly remote metering. Moreover, until now, real-time pricing has had little relevance because (1) very few customers have been capable of adjusting their demand in real-time, and (2) most customers have been covered by contracts or regulated tariffs, and have not been affected by fluctuations in market prices. Real-time pricing will bring new opportunities for *real-time communications* and *control* businesses. Small-scale power production, storage equipment, and demand-altering appliances are new products that retail electricity suppliers can offer to their customers.

The introduction of retail competition increases price risk because wholesale market prices would be available to all customers. To protect themselves against risk, retail customers can create financial arrangements. *Futures contracts*, *options contracts*, *forward contracts*, and other new kinds of power purchase and financial contracts will play an important role in the new retail market.

5.4.3. Retail Access Tariffs

Under retail choice, the customer may buy energy from any retailer or directly from the wholesale market, but the consumer continues to pay a regulated *retail access tariff.* Retail access tariffs include network costs and other costs that the regulatory authorities decide to charge consumers, e.g., energy efficiency and renewable program costs and competition transition charges.

To implement customer choice and retail competition, the regulator must guarantee open access to distribution networks under nondiscriminatory and transparent retail access tariffs. The design of the retail access tariff is critical to ensure equity, financial stability of the system, and the avoidance of uneconomic bypass (customers bypassing the regulated distribution networks).

One method of setting retail access tariffs is to *deduct* costs avoided by the distribution utility when the customer chooses another retailer *from* the full-service tariff. This is called *net-back pricing* because the cost of generation and other costs that are avoided are "netted back" (subtracted) from the normal final tariff to calculate the retail access tariff. In Spain, for instance, retail access tariffs for eligible customers are calculated from the full retail tariffs charged to regulated customers for all electricity services (energy + network + other bundled services + system costs) minus the price of energy. In this way, the following relationship is met:

$$\textit{Retail Access Tariff} = \textit{Full Retail Tariff} - \textit{Wholesale Energy Price} \tag{5.1}$$

such that

1. No discrimination is established between regulated and deregulated customers; all customers contribute to regulated system costs.
2. Coherency is achieved through economic signals, e.g., network and energy prices.
3. A unified settlement procedure is used for regulated tariffs (access and full tariffs).
4. The tariff system takes into account spatial (voltage levels) and time discrimination.

In Argentina, regulated access tariffs are calculated taking into account three types of costs: (1) wholesale capacity and energy costs, (2) network distribution costs, and (3) commercial costs. These costs are allocated to the different categories of regulated customers (small, medium, and large loads) according to their metering equipment and the voltage level. Large users, who buy directly from the wholesale market, pay only the retail access tariff that corresponds to their network distribution costs. Commercial costs are shared only by small and medium-sized customers, and wholesale capacity and energy services are separately contracted (see Chapter 9).

Distribution network costs are likely to become the main component of retail access tariffs. Economic theory suggests that these costs should be allocated among network users so each transaction would be charged the marginal cost it imposes on the system, subject to ensuring adequate total revenues for distribution companies. Studies have shown that marginal distribution costs vary greatly by location and time. Distribution network costs will continue to be allocated through the retail access tariff among different customer classes and for different voltage levels (high-, medium-, and low-voltage networks), taking into account practical considerations such as

1. The number of customers
2. The type of meters installed
3. End-use (to reflect different elasticities of demand and to allocate the difference between marginal cost and total revenue)
4. Bill impacts (from tariff rebalancing)
5. Network topology
6. Economies of scale in system upgrades

5.4.4. Distribution Company Regulation

The two main activities traditionally carried out by distribution companies (when integrated with generation and transmission) have been (1) operation, maintenance, and investment in distribution networks that connect transmission substations to customer loads, and (2) selling electricity at retail to final customers. DISCOs have an obligation to supply all customers with no discrimination in their franchise areas.

Under restructuring and competition, network activities continue to be regulated because they are considered natural monopolies (see Chapter 2). Distribution regulation must ensure adequate revenues to compensate operation and maintenance costs and to remunerate capital investments in new installations. Assessment of "efficient" distribution costs is a difficult task that must be conducted by the regulator to determine required revenues. Distribution regulated costs can be classified as

- Operation and maintenance of network installations
- Investment in network reinforcements and new installations
- Distribution network losses
- Metering, billing, commercial management, and DSM programs for regulated customers

With retail choice, distribution companies no longer have the monopoly to sell energy in their franchise areas. In addition, other traditional distribution activities can be opened to competition. Consequently, if a distribution company is allowed to conduct retail or other activities in competition with other retailers, some sort of separation between the nonregulated and the regulated businesses is required. The regulator typically imposes rules governing the relationship between the regulated distributor and their affiliated retail operations.

The regulation of distribution is evolving from the traditional ROR/COS regulation (where distribution was bundled with the rest of utility activities) to PBR regulation, specifically designed for unbundled network activities. Most of restructuring experiences that have implemented customer choice (e.g., UK, California, Norway, Spain, and Argentina) have adopted PBR schemes to regulate DISCOs (see Chapter 4). Quality control mechanisms, associated with the concern that cost reduction can lead to quality degradation, are being used in most of these PBR schemes. DISCOs continue to supply energy to customers that are still considered captive or customers that have not switched to other suppliers.

EXERCISE 5.1. DETERMINING DISPATCH IN A POOLCO MARKET

Table 5.1 presents the bids that the MO receives from generators to match a projected demand of 1000 MW. Determine which generators would be scheduled and the market-clearing price. (This problem relies on Fox-Penner, 1997, p. 188.)

EXERCISE 5.2. DETERMINING DISPATCH BASED ON PHYSICAL BILATERAL CONTRACTS

Table 5.2 presents the physical quantities contracted between buyers and sellers, with the associated firmness and duration. If firm contracts have dispatch priority over nonfirm contracts, and longer contracts have priority over shorter contracts, determine which transactions would be scheduled by the SO in the next hour when projected demand is 1000 MW. If contract prices would establish the dispatch priority, as in a poolco market, what would be the scheduled contracts and the market-clearing price? For economic efficiency, assuming that the demand is inelastic and the generation market is competitive (contract prices reflect true marginal generation costs), what is the social welfare loss (see Chapter 2) from bilateral trading

Table 5.1. Generation Bids

Generator	Quantity (MW)	Price ($/MWh)
A1	250	10
A2	100	16
A3	100	21
A4	50	30
B1	200	12
B2	50	21
B3	50	24
C1	150	14
C2	100	20
C3	100	25
D1	200	23
E1	50	22

Table 5.2. Bilateral Contracts

Generator (Seller)	Distributor (Buyer)	Quantity (MW)	Firm (F) or Nonfirm (N)	Duration (hours)	Price ($/MWh)
E	c	200	F	18	12
A	g	150	N	24	20
M	a	100	F	24	16
B	f	100	N	4	10
D	d	150	F	48	25
F	a	250	F	48	14
L	k	50	N	16	27
C	e	100	F	out	13
G	h	50	N	4	15

Source: Fox-Penner (1997, p. 192).

when compared to centralized pool trading? (This problem relies on Fox-Penner, 1997, p. 292.)

EXERCISE 5.3. GENERATOR REVENUES AND LONG-RUN CAPACITY

Consider a system with 6000 annual hours of off-peak demand and 2000 hours of peak demand (ignore the remainder of annual hours). The corresponding demand functions are

$$P_{\text{offpeak}} = 25 - (Q/1000)$$

$$P_{\text{peak}} = 65 - (Q/1000)$$

where P is price (US$/MWh) and Q is quantity (MWh).

The generation system consists of two types of generators (in units of 100 MW each):

- Type I Generators: Total capacity is 10,000 MW, marginal cost is $15/MWh, and fixed cost per MW is $70,000/year.
- Type II Generators: Total capacity is 10,000 MW, marginal cost is $22/MWh, and fixed cost per MW is $56,000/year.

5.3.1. Compute the short-run supply and demand equilibrium during off-peak and peak consumption hours.

5.3.2. Calculate the net annual revenues per MW of each generator type: first, assuming that the last generation bid accepted sets the market price; and second, assuming that the demand bid sets the market price.

5.3.3. Calculate what should be the system capacity configuration to ensure the long-run equilibrium if the demand bid sets the market price. Note that in a competitive market with free entry and exit, the *net* annual revenues (after payments for capital and noncapital inputs) for each generator would be equal to zero. Generators are able to exactly cover their fixed costs with their short-run operating profits, so there is no incentive for entry or exit.

EXERCISE 5.4. GENERATOR PROFITS WITH AND WITHOUT A CONTRACT FOR DIFFERENCES

A generator with 150 MW of capacity and \$20/MWh marginal cost is selling its energy by bidding into a poolco market. Last year's market clearing prices were \$50/MWh during peak hours (25% of total period), \$30/MWh during shoulder hours (50% of total period), and \$10/MWh during valley hours (25% of total period).

5.4.1. If the generator's total capacity was available during the entire year, calculate the "variable profits" (revenues minus marginal cost, ignoring fixed costs) that the generator would obtain from energy sales to the pool if it were bidding according to its marginal cost.

5.4.2. Based on this, the generator has made a Contract for Differences (CfD) with a customer for this year agreeing on a constant quantity of 100 MW at a price of \$30/MWh. However, this year, peak and shoulder market prices (because a surplus of hydroelectric resources) have been depressed to \$35/MWh and \$25/MWh, respectively. Compute this year's variable profits. How much would these profits change if the generator had not made the contract?

5.4.3. Considering the same contract, now assume peak and shoulder market prices increase to \$55/MWh and \$35/MWh. Next, compute the generator's variable profits if the available generator capacity has decreased to 75 MW during the whole year. How would these profits change if the generator had not made the contract?

EXERCISE 5.5. THE VALUE OF TRANSMISSION EXPANSION BETWEEN TWO ZONES

Two zones A and B are not electrically connected. Zone A has a demand of 3400 MW and a supply curve that increases \$0.01/MWh for each MWh supplied. The market-clearing price in Zone A is \$34/MWh (the point where supply and demand curves intersect in Figure 5.4). Zone B has a demand of 2300 MW and a supply curve that increases \$0.02/MWh for each MWh supplied. The market-clearing price in Zone B is \$46/MWh (see Figure 5.5). Here, demand is completely inelastic, e.g., where there are no energy bids from the demand side.

A transmission expansion study examines the economic feasibility of an interconnection between A and B. The study's assumptions are

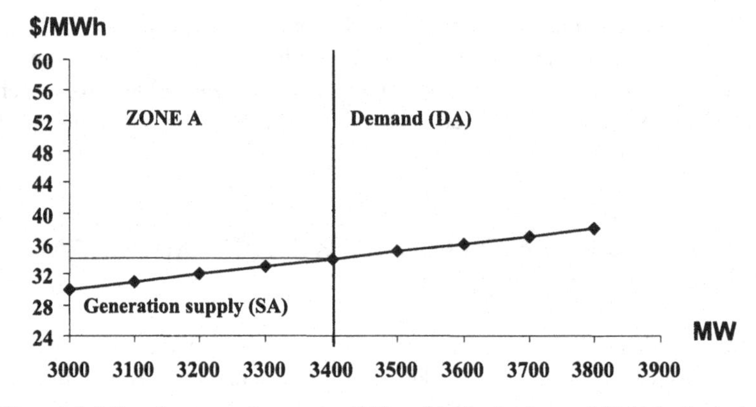

Figure 5.4. Supply generation curve (SA) and inelastic demand (DA) in Zone A.

1. The interconnection will be built with 230 kV lines
2. Each 230 kV line has a transfer capacity of 300 MW and an investment cost of $110,000/km
3. The distance between A and B is 150 km
4. The expected equipment life is equal to the loan life of 30 years
5. The real cost of capital (r) is 10%
6. The interconnection would operate at the maximum transfer capacity during 980 hours per year when the generation and demand conditions as given above in Zones A and B occur

Under these assumptions, the levelized cost per MWh of an interconnection between A and B equals $6/MWh (see Chapter 3, Section 3.2.1 for calculation of the levelized cost of an investment).

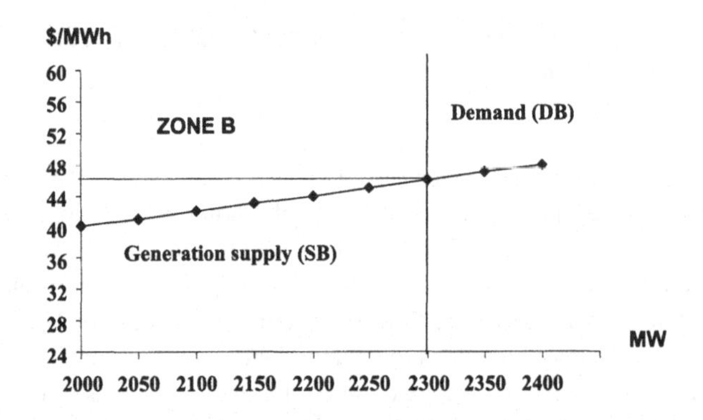

Figure 5.5. Supply generation curve (SB) and inelastic demand (DB) in Zone B.

5.5.1. Determine the optimal transfer capacity for the interconnection.

5.5.2. If only steps of 300 MW are technically feasible for transmission lines between Zones A and B, determine how many lines (of 300 MW each) can be economically justified.

5.5.3. Compute the benefits that market participants (generators A and B and customers A and B) would obtain from the new interconnection. The generator producer surplus is equal to revenues less generation marginal costs given by the supply curve. The consumer surplus is obtained as the difference between what the consumer would have been willing to pay (with inelastic demand this is set to a high value, for example, $250/MWh) and the market price (see Chapter 2 for definition and computation of consumer and producer surplus). Calculate the congestion rents obtained by the interconnection owner. Are congestion rents enough to pay all investment costs? Propose a possible scheme, based on market participant benefits, to allocate the investment costs not recovered as congestion charges.

EXERCISE 5.6. CALCULATE NODAL PRICES IN A THREE-BUS TRANSMISSION SYSTEM

The transmission system represented in Figure 5.6 connects

- Generator 1 with a marginal generation cost of $25/MWh and capacity of 1500 MW
- Generator 2 with a marginal generation cost of $45/MWh and capacity of 750 MW
- Load 3 of 1200 MW

The three lines experience no losses, have equal length and impedance, and have a maximum transfer capacity equal to 1000 MW each. To transmit power from one bus to another, through parallel paths, the flow in each path is inversely proportional to path impedance. That is, if the impedance of one path were double the impedance of the another path, the power flow on the path of lower impedance would be double the flow on the higher impedance path. (Heuristically, the impedance is the

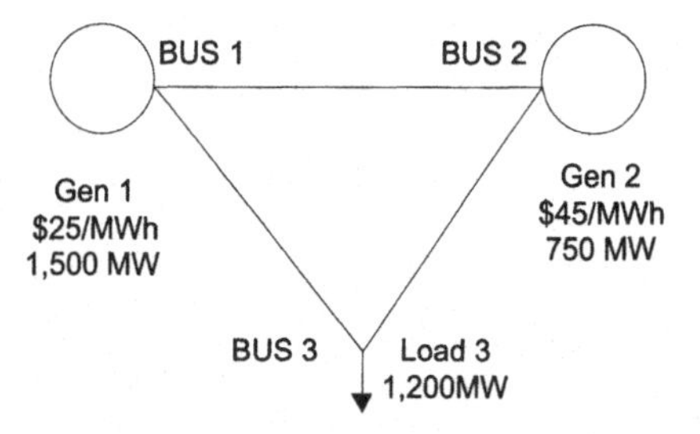

Figure 5.6. Three-bus transmission system.

resistance that a path presents to the power flow.) Calculate the energy prices at the three transmission buses, i.e., the nodal prices, in the following situations. (Remember that nodal prices are calculated in each system bus as the minimum increment of total system marginal costs to supply a demand increment in the particular bus.)

1. All generation capacity is available.
2. Generator 1 capacity decreases to 900 MW due to a forced outage.
3. All generation capacity is available but the transfer capacity of the line between buses 1 and 2 decreases to 300 MW. Also, calculate the congestion charge for the congested path.

CHAPTER 6

THE CALIFORNIA POWER SECTOR

Ryan Wiser, Steven Pickle, and Afzal S. Siddiqui

California offers an instructive case study of electric industry restructuring for a variety of reasons. California has been a leader, not only in electricity market reform efforts in the United States, but also worldwide. It was in California that the development of an independent power industry began in the early 1980s. More recently, California was also among the first U.S. states to open its market to retail competition. It also developed an open wholesale electricity market with both an hourly spot market and a transmission/system operator. California has the longest and most detailed experience with competitive electricity markets in the United States.

Additionally, beginning in the summer of 2000, an electricity crisis gripped the state, resulting in periodic blackouts, unprecedentedly high wholesale electricity prices, and a financial catastrophe for the state's electric utilities. Out of this crisis has emerged a new structure for the electricity sector, one that looks different from the one envisioned at the onset of the state's restructuring process. The impacts of the crisis have been profound. Analysts and politicians throughout the world have used California's electricity restructuring experience to rethink the basic tenets of electricity reform, and to reconsider the trend toward increasingly open wholesale and retail electricity markets.

Most of this case study focuses on the design of California's restructured electricity industry before the crisis. The case study begins with a general overview and description of California's power system. It then describes the regulatory framework that has shaped the structure of California's electricity sector, pre- and postreform. It highlights the design of California's wholesale electricity market, transmission issues, distribution network regulation, and retail competition. Other aspects of California's restructured electric system are also discussed. The case study concludes with a review of early experience with electricity sector reform in California, the nature of the state's electricity crisis, its causes, and its possible impacts on the design of California's future electricity system.

6.1. GENERAL DESCRIPTION OF THE CALIFORNIA POWER SYSTEM

In late 1996, the California state legislature approved legislation that reorganized the state's electricity industry and introduced retail competition for California's electricity consumers starting March 31, 1998. To understand these and related changes in California's electricity sector it is important to begin by understanding the basic structure of the electricity system in the state.

More than 3000 electric utilities operate in the US to provide electricity service to customers. At the end of 1999, the *net* generating capacity of the US electric power industry stood at more than 779 gigawatts (GW) (EIA, 1999a). (*Net* generation excludes self-generation units and internal generation uses.) Sales to ultimate customers in 1998 exceeded 3240 terawatt-hours (TWh) at a total cost of more than US$218 billion (EIA, 1999a). Although most of the utilities in the US are publicly or cooperatively owned, sales by the 239 private, *Investor-Owned Utilities* (IOUs) represent approximately 75% of total electricity sales (EIA, 1999b).

California is the most populous US state, and generation serving California represents approximately 7% of total generation in the US (EIA, 1998). Before the restructuring of the industry in 1998, California's electric industry consisted of both public and private vertically integrated electric utilities that managed and operated generation, transmission, and distribution systems in the state. The three largest private IOUs are Pacific Gas and Electric Company (PG&E), Southern California Edison (SCE), and San Diego Gas and Electric Company (SDG&E). These three utilities combined have historically served approximately 75% of all load in California. The remainder of the load has been served by a mix of more than 40 smaller investor-owned, government-owned, and cooperative utilities; the largest of these are the Los Angeles Department of Water and Power (LADWP) and the Sacramento Municipal Utility District (SMUD).

California's IOUs are regulated by

1. The California Public Utilities Commission (CPUC), which has historically overseen utility rates and operations
2. The California Energy Commission (CEC), which oversees new plant siting and construction
3. The Federal Energy Regulatory Commission (FERC), which regulates wholesale electricity trade and interstate transmission

Publicly and cooperatively owned utilities are regulated by municipal, county, and other oversight bodies, subject to state and federal law.

Each of the main IOUs was, historically, responsible for matching load and resources to maintain electrical reliability and to match scheduled and actual flows at tie points, where the utilities are connected to other power producers. Each utility, having an obligation to serve load within its service territory, developed its own generation and demand forecasts, operated generating plants, and entered procurement contracts for the fuel used to generate electricity. Each utility also participated in short- and long-term bilateral contracts for electric power.

6.1.1. Generation

Within California, electricity is generated by more than 1300 power plants. Total electric generation serving California in 1999 was 259 TWh (including self-generation). Historically, about 20% of the electricity used in California has been imported from outside the state (approximately half of which is from coal-burning plants and the other half is hydroelectric).

Until the late 1970s, most electric generating capacity in California was owned by electric utilities and government agencies. Since the late 1970s, however, wholesale competition in electricity generation has been allowed (as discussed in Section 6.2, below). A significant fraction of the total generation serving California now comes from independent, nonutility generators selling to California utilities. For example, in 1996, before electricity restructuring, 81% of California's generating capacity was utility owned; nonutility generators owned the remaining 19%. After the utility divestiture of generation beginning in 1998 and stimulated by electricity reform, utility ownership dropped to 46% of the total, compared to 54% for nonutility generators. Most of the nonutility owned generation comes from natural gas and renewable energy.

Table 6.1 presents the resource composition of the electric generation serving California, including in-state electricity generation and imports. California's electricity supply is diverse, with substantial amounts of gas, hydroelectric, coal, nuclear, and renewable energy generation.

6.1.2. Transmission and Interconnections

The electric utilities in California are linked through an extensive network of transmission lines. In California, the main transmission grid consists of 500 kV, some

Table 6.1. Electricity Production by Generation Resource (Including Imports), 1999

Resource	Energy
Hydroelectric	20.1%
Nuclear	16.2%
Coal	19.8%
Gas and oil	31.0%
Renewables and other	12.2%
Total	259 TWh

Source: www.energy.ca.gov/electricity/sb1305/index.html.

230 kV, and 500 kV DC high-voltage transmission lines. Some larger customers receive service at these high voltage levels. For smaller customers, the voltage is stepped down to lower voltages (e.g., 120 volts for residential customers).

The three main investor-owned utilities—PG&E, SCE, and SDG&E—have historically owned and operated the bulk of the transmission grid in California (some smaller utilities also own pieces of the transmission system). These same utilities also served as managers of the coordinated operation of the generation and transmission systems, performing economic and technical functions such as security analysis, economic dispatch, unit commitment, etc. Under restructuring, these IOUs were to maintain ownership of their transmission assets and responsibility for their maintenance. But as discussed in more detail below, an *Independent System Operator* (ISO) was created to operate the bulk of the transmission system in the state.

The state of California also has many transmission interconnections with adjacent states. This allows for power transfers from throughout the western United States. The most important of these transfers are with the Northwest and Southwest. Of the 18% of electricity that came from imports into the state in 1999, the CEC estimates that 53% came from the Northwest and 47% from the Southwest interconnections.

6.1.3. Distribution

Distribution power lines generally include line voltages at and below 50 kV. The distribution networks in California continue to be owned and operated by the various utilities in the state, including the three large IOUs and the smaller investor-owned and government-owned utilities. Under restructuring, these companies are continuing to provide distribution services to all electric customers within their respective service territories.

6.1.4. Consumption

Table 6.2 shows the electricity consumption and number of customers in each customer class in California. Regarding electricity consumed in California in 1998, Table 6.3 shows the breakdown by utility service territory. In total, 75% of total

Table 6.2. Electricity Consumption and Number of Customers in California, 1998

Customer Class	Customers	Consumption (TWh)
Residential	11,331,398	75
Commercial	1,522,665	86
Industrial	39,902	59
Other	47,000	7
Total	12,885,000	226

Source: EIA (1999c).

Table 6.3. Electricity Consumption by Utility Service Territory, 1998

Utility	Consumption (TWh)	Percent of Total
Southern California Edison	76.3	33.8%
Pacific Gas & Electric	75.7	33.5%
Los Angeles Department of Water & Power	21.7	9.6%
San Diego Gas & Electric	16.3	7.2%
Sacramento Municipal Utility District	9.1	4.0%
Other	26.9	11.9%
Total	226	100%

Source: www.eia.doe.gov/cneaf/electricity/esr/t17a.txt.

consumption comes from investor-owned utilities and 25% from government-owned utilities.

6.1.5. Concentration Levels

The large IOUs continue to own the bulk of the transmission and distribution networks in California. Retail competition allowed customers to select alternative energy service providers. As a practical matter, however, most of the customers who were eligible for retail competition remained with the incumbent utilities before retail competition was discontinued. Because of the electricity reform process, the three major IOUs were required temporarily to sell into and purchase from a centralized *Power Exchange* (PX). This was to reduce the potential for the abuse of horizontal market power. Further, the ISO was created to reduce the likelihood of vertical market power. To reduce concentration levels in electricity generation, the three major IOUs completed the sale of much of their California fossil-fuel-fired generating assets, as well as some of their other generating plants. As we discuss later with respect to the energy crisis in California, however, even these mitigation measures have been inadequate to protect California's electricity system from the exercise of market power.

6.1.6. Plant Investment

Total plant investment by investor-owned utilities in the United States, based on information provided to the Energy Information Administration (EIA), is listed in Table 6.4. As this table shows, most of the plant investment comes from electricity production and distribution.

6.1.7. Electricity Prices

Table 6.5 shows California's 1998 electricity rates, by customer class, and compares them to U.S. average rates. Rates in California are clearly higher than the na-

Table 6.4. Net Electric Utility Plant Investment in the United States, 1996

Investment Type	Percent of Total Investment
Electricity production	54.6%
Transmission	11.6%
Distribution	28.8%
Other	5.0%

Source: EIA (1997).

tional average; this has been a key driver in the state's restructuring efforts. Moreover, the year 2001 ushered in significant increases in these already high rates as a result of the state's electricity crisis.

6.1.8. Economic and Energy Indices

California's economic and energy indices may be summarized as follows (source: www.eia.doe.gov/emeu/states/main_ca.html):

- Population (2000): 33.871 million
- Size: 0.4 million square kilometers
- Gross State Product (2000): $1229B (1996 U.S.$)
- Inflation rate (2000): 4.0%
- Oil production (2000): 741,000 barrels per day (bbl/day)
- Natural gas production (2001): 0.344 trillion cubic feet (Tcf)
- Electricity consumption (2000): 221 million MWh
- Energy consumption per capita (2000): 253 MBtu (second lowest among states in the United States)

Table 6.5. Electricity Rates in California and the United States ($/MWh), 1998

Customer Class	California Electricity Rates	U.S. Average Electricity Rates
Residential	105	83
Commercial	97	74
Industrial	63	45
Other	75	68
Average	90	68

Source: EIA (1999d).

6.2. THE NEW REGULATORY FRAMEWORK

Although California was one of the first U.S. states to inject competition into its electricity market, many changes enacted in the state emerged from a series of broader regulatory changes at the national level. In the United States, the federal government regulates *interstate* commercial transactions and the states have authority to regulate *intrastate* commerce. In the electricity industry, this division of regulatory power has typically meant that the federal government (through the Federal Energy Regulatory Commission, or FERC) has overseen wholesale electricity transactions and issues related to transmission pricing and access. The state commissions (frequently termed public utility, public service, or corporation commissions) have regulated retail electricity transactions and access to the distribution grid. To understand California's new market structure and regulatory system, it is necessary to first introduce key federal legislation and regulations.

6.2.1. U.S. Federal Legislation and Regulation

6.2.1.1. Public Utility Holding Company Act of 1935. The current structure of the U.S. electric power industry was established with the passage of the Public Utility Holding Company Act (PUHCA) in 1935. PUHCA's purpose was to break up the large, and essentially unconstrained, holding companies that then controlled much of the country's electric and gas distribution networks. Under the Act, the Securities and Exchange Commission (SEC) was given the power to break up interstate utility holding companies by requiring them to divest their holdings until each became a single consolidated system serving a circumscribed geographic area. The law further required electricity companies to engage only in business activities essential and appropriate for the operation of a single, vertically integrated utility. This restriction virtually eliminated the participation of nonutilities in wholesale electric power sales. PUHCA also required that any utility engaging in interstate electricity trading or transmission be regulated by the Federal Power Commission, which, in 1977, became FERC (EIA, 1993, 1996).

6.2.1.2. Public Utility Regulatory Policies Act of 1978. The landscape created by PUHCA remained largely intact until 1978 when, spurred by increased concern over U.S. dependence on foreign oil in the wake of the OPEC oil embargo and by increased environmental awareness, the U.S. Congress passed the Public Utility Regulatory Policies Act (PURPA). Designed to promote energy efficiency and increase cogeneration and renewables, PURPA was significant for the utility industry because it ensured a market for nonutility generated electricity.

PURPA established a new category of independent electricity generators called *qualifying facilities* (QFs). These were defined as nonutility power wholesalers that were either (1) cogenerators or (2) small power producers using specified renewable energy resources. (Eligible renewable resources included biomass, waste, geothermal, solar, wind, and hydroelectric power under 30 MW.) Under the law, utilities were required to purchase whatever electricity was offered from any QF at a rate equal to the purchasing utility's incremental or avoided cost of production (PURPA, 1978; EIA, 1996; Watkiss and Smith, 1993). Although competitive bidding has been used to set avoided cost payments in recent years, many early QF contracts were set at high, fixed avoided cost levels. (This is discussed further below in the section on stranded cost issues.) PURPA served as a first step in opening the wholesale electricity market to competition and allowing participation by nonutility entities.

6.2.1.3. Energy Policy Act of 1992. The introduction of wholesale competition was furthered with the passage of the Energy Policy Act (EPAct) in 1992. As with PURPA, concern about America's oil dependence—heightened by the 1991 war with Iraq—was a key driver of the new energy regulations outlined in EPAct. Provisions dealing with the utility industry were also driven (1) by an increased awareness of the benefits of new, more decentralized generation technologies, and (2) by an increased sense in academic and policy communities that economic efficiency gains could be realized through the introduction of competition into electricity markets (Joskow and Schmalensee, 1983; Kahn, 1988; Hausker, 1993; Watkiss and Smith, 1993; Joskow, 1997).

In brief, EPAct substantially reformed PUHCA and made it easier for a broader array of nonutility generators to enter the wholesale electricity market by exempting them from PUHCA constraints. The law made these changes by creating a new category of power producers called *Exempt Wholesale Generators* (EWGs). EWGs differ from PURPA QFs in two ways. First, EWGs are not required to meet PURPA's cogeneration or renewable fuels limitations. Second, utilities are not required to purchase power from EWGs. Instead, marketing of EWG power is facilitated by provisions in EPAct that give FERC the authority to order utilities to provide access to their transmission systems on nondiscriminatory terms. Specifically, EPAct instructs FERC to require utilities to make transmission service available at "just and reasonable" rates, designed to cover all "legitimate, verifiable, and economic costs," subject to the condition that any incremental costs be recovered from the entity seeking transmission service and not from the transmitting utilities' existing customers (EPAct, 1992; Hausker, 1993; Watkiss and Smith, 1993; EIA, 1996).

These transmission provisions paved the way for open-access wholesale electricity transactions on a national basis. This new market has been especially significant for smaller, transmission-constrained utilities that are no longer dependent on adjacent utilities for wholesale power. Whereas EPAct established a legal framework for widespread wholesale competition, the specifics of the wholesale market were laid out later in FERC Orders 888 and 889.

6.2.1.4. FERC Orders 888 and 889. Following EPAct, FERC began to review and mandate wholesale transmission requests on a case-by-case basis beginning in 1993. (FERC's first ruling mandating wholesale transmission access came in October 1993 in *Florida Municipal Power Agency v. Florida Power & Light Co.*) While hearing case-by-case requests, FERC was also looking for a more comprehensive way to mandate terms for open-access wholesale electricity transactions.

After issuing an initial set of proposed rules in a document known as the "Mega-NOPR" and taking comments, FERC released two major rules on transmission access in Spring 1996. (NOPR stands for "notice of proposed rulemaking.") These rules, termed Orders 888 and 889, spelled out the specific details and requirements for wholesale electricity transactions and established a real-time, transparent trading communications system. In brief, Order 888

- Requires all utilities under FERC jurisdiction to file nondiscriminatory open access transmission tariffs, available to all wholesale buyers and sellers of electricity
- Requires utilities to take service under their filed tariff rates for their own wholesale electricity purchases and sales
- Allows utilities to recover all legitimate, prudent, and verifiable but now uneconomic (i.e., "stranded") wholesale costs and investments incurred before July 1994

(We discuss stranded costs—a key issue in the US debate over utility industry restructuring—in Section 6.6.)

The first two of these provisions were intended to enable all electricity providers to have access to the transmission grid on equal terms for both point-to-point and network transmission services, including ancillary services. The final provision recognized the legitimacy of utility concerns regarding the recovery of costs incurred under a different regulatory regime and with different expectations about the likelihood of cost recovery. Finally, Order 888 required that municipal and other utilities not under FERC jurisdiction nonetheless provide reciprocity should they wish to avail themselves of the open access tariffs offered by utilities complying with the FERC order. In other words, the reciprocity rule ensures that if a municipal utility wanted to purchase transmission service from an IOU under the terms of Order 888, they must offer service on similar terms (FERC, 1996a; EIA, 1996).

On the same day it issued Order 888 FERC also issued Order 889. This compelled utilities to create and use an *Open Access Same-time Information System* (OASIS) providing all transmission customers with standardized electronic information on transmission capacity, prices, and other essential market information. The rule also requires that transmission operations personnel at utilities function independently of generation and wholesale trading personnel. Finally, although not mandating the creation of independent system operators (ISOs), Order 889 encourages the creation of ISOs and recognizes that they would fall under FERC's jurisdiction (FERC, 1996b; EIA, 1996). Subsequent orders have clarified FERC's position on the formation, design, and governance of regional transmission organizations and ISOs.

6.2.2. California State Regulation and Legislation

6.2.2.1. California Public Utilities Commission Activity. As federal legislation and regulations began to restructure the US electricity industry, regulatory officials in California started to investigate the possibility of even more sweeping changes. In the early 1990s, the California Public Utilities Commission (CPUC) began to explore the possibility of introducing some form of retail competition within the state's three investor-owned and CPUC-regulated utilities. California was emerging from an economic recession. At the same time, the state's electricity prices were more than 40% higher than the national average (even for industrial customers) and as much as twice those in neighboring states (EIA, 1995).

Concerned about the loss of businesses seeking cheaper electric rates and aware of successful deregulation and cost savings in other industries, the CPUC issued its *Yellow Paper* in February 1993. It detailed options for introducing retail competition in electricity service (CPUC, 1993; Pickle, Marnay, and Olken, 1997). Besides competitiveness concerns and deregulation experience in other industries (e.g., telecommunications, natural gas, and transportation), drivers behind the move toward retail competition in California also included (1) an increased perception of the inefficiencies of traditional ROR regulation, (2) a desire to tap new and potentially more cost effective generation technologies, and (3) advances in communications and information technologies necessary for price discovery and the unbundling of various electricity-related services (see Joskow and Schmalensee, 1983; EIA, 1996; Pickle, Marnay, and Olken, 1997).

In April 1994 the CPUC released a follow-up document called the *Blue Book*. The *Blue Book* advanced a more detailed proposal for retail competition in the service territories of the three major California IOUs. Specifically, the *Blue Book* called for (Blumstein and Bushnell, 1994; CPUC, 1994):

- The introduction of a comprehensive wholesale spot market trading system (e.g., similar to a UK-style poolco)
- The introduction of (1) a nonbypassable *Competition Transition Charge* (CTC) designed for the recovery of utility stranded costs and assets and (2) another similar, but much smaller charge, to fund public-purpose programs (discussed in Section 6.6)
- The use of performance-based ratemaking (PBR) in place of traditional ROR regulation for transmission and distribution
- A gradual phase-in of retail competition and direct access beginning in 1996 and ending in 2002.

The CPUC made clear that the *Blue Book* was proposing retail competition for generation only. "Transmission and distribution services," wrote the Commission, "as well as system control and coordination services, will continue to receive regulatory oversight" (CPUC, 1994, p. 31). The regulation and general pricing guidelines for transmission and distribution services are discussed further in Sections 6.4 and 6.5.

The primary purpose of the *Blue Book* was to elicit comment on the CPUC's proposals. After taking comment, the CPUC issued its decision on the introduction of retail competition in December 1995 (CPUC, 1995). The primary difference between the *Blue Book* and the CPUC's final decision was the CPUC's full adoption of a dual bilateral and poolco system; the decision established a structure that would give consumers the option of conducting bilateral trades, pool-based trades, or both (Bushnell and Oren, 1997).

To create this hybrid structure, the decision called for the creation of two new entities, a PX designed to serve as a clearinghouse or spot market for supply and demand bidding, and an ISO charged with managing grid operations (CPUC, 1995; Bushnell and Oren, 1997). Although the CPUC's decision established the structure for the new California market, it quickly became clear that the scope of the changes envisioned in the decision was greater than the CPUC could mandate on its own. Retail competition in California would require legislative action by the California State Assembly and Senate.

6.2.2.2. Legislative Activity and AB 1890. After a series of hearings and debates, the California State Legislature endorsed the CPUC's proposed market structure with the passage of Assembly Bill 1890 (AB 1890) in August 1996. Signed by the Governor on September 23, 1996, AB 1890 provided the legal basis for competition for electricity service in California. In brief, AB 1890

- Called for the establishment of a PX and an ISO as independent, public-benefit, nonprofit market institutions to be overseen by a five-member Oversight Board
- Required California's IOUs to commit control of their transmission facilities to the ISO
- Allowed for direct, bilateral electricity trading
- Called for a transition to retail competition beginning January 1, 1998, and to be completed by March 31, 2002
- Called for a cumulative 20% rate reduction by 2002 (beginning at 10% in 1998) below 1996 rates for residential and small commercial customers with financing for the reduction to be "securitized" by the issuance of nonrecourse state bonds designed to allow the consolidation of specific utility obligations at lower interest rates and to be repaid by all residential customers with a nonbypassable charge
- Permitted up to 100% of utility stranded costs to be recovered by a nonbypassable CTC, provided it could be collected under the rate cap during the transition period
- Established a separate charge to pay for public-purpose programs designed to support (1) low-income ratepayer assistance and (2) energy efficiency activities, research and development, and renewable energy during the transition period
- Encouraged the divestiture of generation assets by the state's two largest IOUs (PG&E and SCE) to mitigate the abuse of market power
- Encouraged, but did not require, government-owned utilities to offer direct

access, but did require government-owned utilities to offer reciprocity if seeking to sell power into new retail markets in California
- Established functional separation of generation, transmission, and distribution, and rules for unregulated utility affiliates

AB 1890 gave the CPUC authority to issue clarifying orders detailing the rules and procedures required to initiate competition in California. Key among these decisions was the CPUC's determination to allow all IOU customers access to the new market simultaneously on January 1, 1998 (CPUC, 1997a). The CPUC also ruled that competition should be allowed in the provision of metering and billing services (CPUC, 1997b).

Just weeks before the scheduled January 1998 market start date, the CPUC was forced to postpone the opening of the market due to difficulties associated with installing and testing new computer systems. Because of this delay, California's competitive market for electricity opened three months late on March 31, 1998. On that date, all retail customers served by the state's three major IOUs became eligible to take service from new power providers. Approximately two years later, during the summer of 2000, the state's experiences with electricity sector reform degraded into crisis.

Table 6.6 reviews the chronology of key legislation and regulatory decisions discussed in this section. In the next four sections, we discuss in more detail the implementation of AB 1890 and the specifics of the new market design and structure. The final section discusses the early experience with electricity reform and the crisis that followed.

6.3. THE WHOLESALE ELECTRICITY MARKET AND INSTITUTIONS IN CALIFORNIA

In this section, we discuss the wholesale side of the market. In particular, we discuss two key institutions (and several related players) that were created by AB 1890 for wholesale electricity trading in the new California market:

1. A wholesale spot market, the California Power Exchange (PX)
2. The new grid operator, the California Independent System Operator (ISO)

One of these institutions, the PX, ceased operations in early 2001 due to the electricity crisis. Figure 6.1 illustrates the roles of and relationships between key players on both the wholesale and retail sides of the new California electricity market created by AB 1890.

6.3.1. The Power Exchange

Located in Alhambra, California, the California PX was created as a nonprofit corporation. Its primary purpose was to provide an efficient, short-term competitive

Table 6.6. Chronology of Key Legislative and Regulatory Developments

Date	Legislation, Ruling, or Event	Significance
1935	PUHCA passed	Interstate holdings restricted, monopoly service territories required, wholesale market restricted to utilities only, wholesale trading regulated by FPC (later FERC).
1978	PURPA passed	QFs established; utilities compelled to purchase QF energy.
1992	EPAct passed	EWGs established; nondiscriminatory transmission access required.
1993	CPUC "Yellow Paper"	CPUC investigates possibility of instituting retail competition.
1994	CPUC "Blue Book"	CPUC proposal for introduction of retail competition in California.
1995 (March)	FERC issues "Mega-NOPR"	FERC's initial proposal for uniform open access transmission rules and stranded cost recovery.
1995 (December)	CPUC issues D.95-12-063	CPUC's decision calling for the introduction of competition in California's IOU operated electricity market via a hybrid pool and bilateral system.
1996 (April)	FERC issues Orders 888 and 889	Orders require utilities to file and use open access transmission tariffs; allows for wholesale stranded cost recovery; calls for creation of an open access same-time information system (OASIS) for electricity trading.
1996 (August)	AB 1890 passed and subsequently signed into law	California law formally mandating competition for the state's IOUs and establishing the terms for full competition statewide.
1997	AB 1890 implementation rulings	In accordance with AB 1890, CPUC issues rulings spelling out procedures for instituting competition; eliminates customer phase-in; extends competition to meter and billing services.
1998 (March)	California market opens March 31	After a three-month delay caused by computer glitches, California's new electricity market formally opens.
Summer 2000	Electricity crisis begins	The beginnings of the electricity crisis: high wholesale power prices, degraded electricity reliability, and financial losses for the state electric utilities.

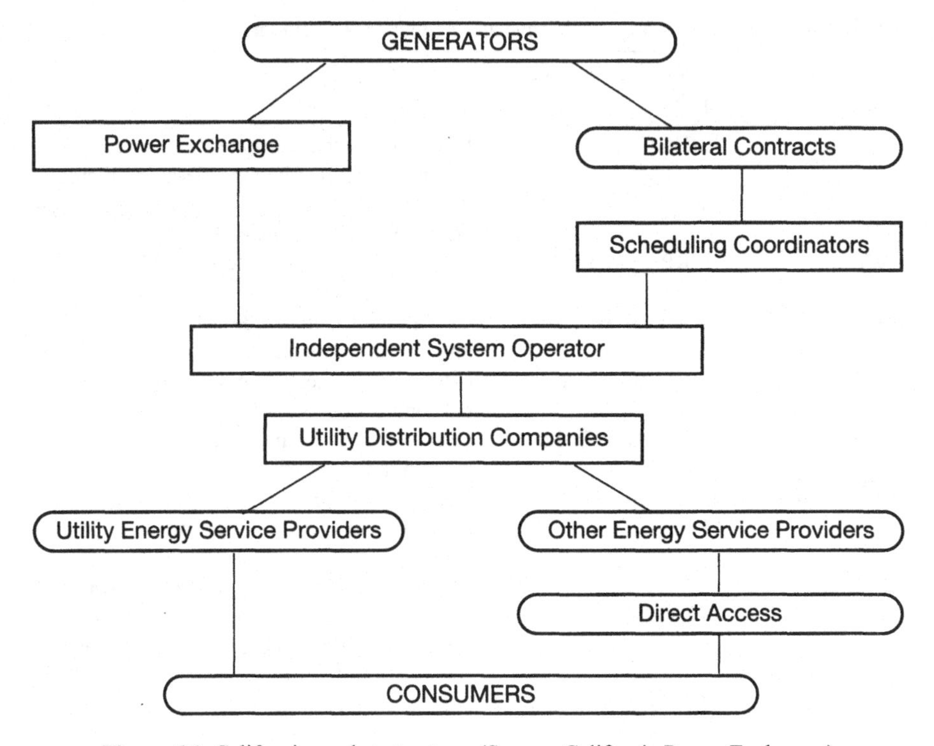

Figure 6.1. California market structure. (Source: California Power Exchange.)

energy market that met the loads of PX customers at market prices. The PX was one of a potentially unlimited number of *scheduling coordinators* authorized to submit balanced schedules and other information to the ISO, which would then conduct real-time dispatch. The PX was operational from the beginning of 1998 through early 2001, at which point it ceased operations. During its period of operation, however, the PX was the most significant player in wholesale electricity trade in the state.

Key features of the PX included the following (California Power Exchange, 1998):

- The PX was open on a nondiscriminatory basis to all suppliers and purchasers.
- The PX calculated the price of electricity hourly for the day-ahead and hour-ahead (later, day-of) markets, according to the supply and demand bids submitted by PX participants.
- PG&E, SCE, and SDG&E, which together represent approximately 75% of the electricity sold in California, were initially required to buy all their electricity from and sell all their generation through the PX. This requirement,

which was implicated as a significant cause of the electricity crisis, was eliminated in 2001.

To participate in the PX market, a prospective participant was required to meet many eligibility requirements, including credit worthiness, identifying metered entities served, etc. Once certified, a participant was allowed to trade in the 24 hourly periods for next-day delivery in the day-ahead market and in the single-hour period for the hour-ahead (and later, day-of) market. Each trade incurred a mutual obligation for payment between the PX and its market participants (Moore and Anderson, 1997). We touch on the operations of these two markets here and also briefly describe the *block-forward market* developed by the PX.

6.3.1.1. The Day-Ahead Market. Procedures for trading in the day-ahead market were as follows (California Power Exchange, 1998; Moore and Anderson, 1997):

- For each hour of the 24-hour scheduling day, participants submitted supply/demand bids to the PX.
- Once bids were received, the PX validated them. Validation consisted of (1) verifying that the content of the bid complied with the requirements of the bid format and (2) checking for consistency with data contained in a master file.
- Once the bids were validated, the PX constructed aggregate supply and demand curves from all bids to set a *market-clearing price* (MCP) for each hour of the 24-hour scheduling day. The MCP was set at the intersection of supply and demand.
- The PX also determined if the submitted bids could create a potential overgeneration condition. If a potential overgeneration condition occurred, the PX was required to inform the ISO. The PX had rules to follow to resolve overgeneration when it occurred.
- Bids initially submitted into the day-ahead auction did not have to be attributed to any particular unit or physical scheduling plant. Such a bid is a *portfolio bid.*
- Portfolio bids that were accepted into the day-ahead market were then broken down into generation-unit schedules that were submitted to the ISO along with adjustment bids (to relieve congestion) and ancillary service bids.
- The ISO then determined, based on all unit-specific supply bids and location-specific demand bids, whether there would be congestion. If there might be congestion, the ISO used adjustment bids to submit an adjusted schedule to the PX and other scheduling coordinators.
- These adjusted schedules and ISO-determined usage charges became the foundation for zonal MCPs (discussed below) and the final schedule submitted to the ISO.

- Schedules could consist of imports, exports, transfers, or generation. Generator schedules were modified to compensate for transmission losses.

6.3.1.2. The Hour-Ahead/Day-Of Market. This market originally began as an *hour-ahead market* but was reconfigured in January 1999 to a *day-of market* to accommodate market participants. In the original hour-ahead market, bids were submitted to the PX at least 2 hours before the hour of operation. These were unit-specific bids; portfolio bids were not allowed. The purpose of this market was to give participants an opportunity to make adjustments based on their day-ahead schedules so that they could minimize real-time imbalances. The MCP was determined the same way as the day-ahead market. The PX announced price and traded quantities to PX participants immediately after the hour-ahead market was closed.

Due to the lack of activity in the hour-ahead market, however, the PX introduced the day-of market. The day-of market was similar in some respects to the earlier hour-ahead market, but conducted its 24 hourly auctions during three auction periods at 6 a.m., noon, and 4 p.m. Auction period prices became available throughout the day.

6.3.1.3. Block-Forwards Market. The California PX also developed a *block-forwards market* intended to offer price hedging services. The block-forwards market offered participants standardized contracts for on-peak energy on a forward month basis. Each contract was based on a specific future month at a specific quantity for the 16-hour on-peak period, from 6 a.m. to 10 p.m., excluding Sundays and designated holidays. Trading was conducted between 6 and 10 a.m. weekdays, and prices were posted publicly at 1 p.m. on trading days. Electricity was required to be delivered to a specific California point. Essentially, the block-forward market was a means for market participants to hedge against price volatility in day-ahead trading (California Power Exchange, 1999).

6.3.2. The Independent System Operator

Located in Folsom, California, and charged with ensuring open access and maintaining the reliability of the transmission grid, the ISO

1. Coordinates day-ahead and hour-ahead schedules from all schedule coordinators and determines adjustments to relieve congestion
2. Buys and provides ancillary services as required
3. Controls the dispatch of generation
4. Performs real time balancing of load and generation

We touch briefly on these tasks, beginning with congestion management.

6.3.2.1. Congestion Management. The ISO manages transmission congestion. The price of transmission services is based on marginal-cost, locational pricing through zonal transmission usage charges (Bushnell and Oren, 1997; Moore and

Anderson, 1997; Shirmohammadi and Gribik, 1998). Rather than detail the ISO's congestion management role here, we have included a discussion of congestion issues in the following section that covers transmission pricing generally.

6.3.2.2. Ancillary Services. In its role of ensuring electricity reliability, the ISO oversees an ancillary services market and schedules ancillary services that have been provided by scheduling coordinators. These ancillary services include (1) automatic generation control, (2) spinning reserve, (3) nonspinning reserve, (4) replacement reserve, (5) reactive power, and (6) black-start generation. (This last service is also known as "startup service" and consists of generating units that can provide energy to the network without any outside electricity.) The ISO holds hourly auctions for the first four services, and purchases reactive power and black-start generation under long-term contracts. (These services are provided by generating plants with *Reliability Must Run*, RMR, contracts.)

Considerable attention has been paid to the operational experience and problems encountered in the ancillary services market (Gómez, Marnay, Siddiqui, Liew, and Khavkin, 1999). Although the ancillary services auction process is designed to be competitive, due to a lack of bidders, the ISO imposed price caps to prevent market participants from using their market power to increase prices (Wolak, Nordhaus, and Shapiro, 1998).

6.3.2.3. The Real-Time Market. When it comes time to conduct actual dispatch, the ISO uses a real-time market to adjust power generation to match load in real time. This process is conducted using bids for supplemental energy (i.e., capacity that has not been scheduled) and the generating units providing ancillary services. The ISO sorts the bids by price into merit order and calls upon the bidders when it is necessary to adjust the balance between generation and load. The last unit called upon in each ten-minute trading period defines the equilibrium price in the real-time market.

Real-time imbalances result from differences between scheduled and metered values for supply and demand. When meter data are processed, the imbalance for each hour in each zone is calculated as the difference between the participant's use of power resources (generation and purchase contacts) and power commitments (sale contracts and consumption). Participants are charged for the difference between actual and scheduled load based on the price in the real-time market. Although this market was intended to only address small imbalances between supply and demand in real time, immediately after the PX ceased operations, a considerable amount of electricity trading was taking place in the real-time market.

6.3.3. Bilateral Trading

Since the inception of retail competition in 1998, wholesale electricity trading in California could have been conducted either in the PX (during its existence) or through bilateral agreements (or through the ISO in the real-time market). In all cases, trades must be scheduled with the ISO. As such, the chief difference between bilateral and PX trading lies in which scheduling coordinator provides the required scheduling information. As with the PX, independent scheduling coordinators of bilateral trades must provide the ISO with balanced schedules and settlement-ready meter data. Independent scheduling coordinators can aggregate supply and demand bids and could effectively compete with the PX (before its closure).

Further, as in other wholesale electricity markets, buyers and sellers have the option of engaging in financial rather than, or in addition to, physical trades. Although we do not discuss these types of agreements in detail here, financial arrangements take the form of exchange-based futures, options contracts, and nonexchange forward agreements, including contracts-for-differences (CfDs). The New York Mercantile Exchange (NYMEX) has offered a number of futures and options contracts for delivery at the California–Oregon border, at the Palo Verde interchange (in southeast California), and at other locations. For a review of these and other financial instruments used to hedge risk in wholesale electricity trading, see Stoft, Belden, Goldman, and Pickle (1998).

6.4. TRANSMISSION ACCESS, PRICING, AND INVESTMENT

Before the development of retail competition in California, limited wholesale competition for electricity generation existed. Within this system, EPAct and FERC required open, nondiscriminatory access to the transmission system, as discussed in Section 6.2. Typically, each integrated electric utility imposed transmission charges on generators through capacity-based, firm, take-or-pay contracts. Each of the three major IOUs in California was responsible for the reliable operation of their transmission systems. Transmission congestion was managed internally by each utility and through coordination among the various utilities. Explicit congestion charges were not collected, although redispatch costs were effectively included in tariffs. Transmission investment and planning were managed by individual utilities and regional transmission groups, and overseen by state and federal utility regulators.

Electricity restructuring in California required new structures for transmission access, pricing, investment, and regulation, much of which was developed through a process of negotiation and compromise. The structure itself emerged from

1. The CPUC's December 1995 decision
2. The state's restructuring legislation (AB 1890)
3. The subsequent FERC filings by the IOUs (WEPEX, 1996; ISO/PX, 1997)
4. The ISO operating agreement and tariff (ISO, 1998)

Under restructuring, the ISO has responsibility for operating the transmission grid of the three large IOUs and, if necessary, rationing access to congested paths. The ISO provides transmission access on a nondiscriminatory basis to all parties. The IOUs, however, continue to own the transmission assets, and earn a regulated rate of return on those assets. As described by Bushnell and Oren (1997), there are three types of transmission charges in California:

1. *Transmission access charges* to recover the sunk costs of transmission investment
2. *Congestion charges* to reflect the operational costs of transmission congestion
3. *Loss compensation* to reflect the operational costs of using the grid

6.4.1. Access Charges

The access charges, or network tariffs, were designed to recover the full revenue requirements (i.e., all network and investment costs) of the transmission facilities transferred to the ISO's operational control by each transmission owner (primarily the three IOUs, although government-owned utilities could also join). The access charges are levied on all end-use customers withdrawing energy from the ISO-controlled grid. They are designed as a single, rolled-in rate that is uniform for similar customers in each utility's service area.

Actual rates and allocation methods were, at least initially, established by the CPUC for the major IOUs. Although it has been controversial, a major attraction of this form of cost allocation is the minimization of cost-shifting across utilities and between customers of each existing utility. To help overcome the free-rider problem, utilities found to be "dependent" on the transmission assets of another utility are responsible for paying some of the revenue requirement of that utility's transmission assets.

6.4.2. Transmission Congestion Charges

Transmission congestion occurs whenever power deliveries are limited by the size or availability of transmission resources needed to serve load. The purpose of congestion management is to allocate the use of, and determine the marginal value of, constrained transmission lines. In its pure form, to alleviate congestion, locational prices should be defined at every bus of the network through "nodal" pricing (see Chapter 5, this volume; Schweppe, Caramanis, Tabors, and Bohn, 1988).

To simplify this approach, buses in California have thus far been combined into pricing "zones"; a zone is a part of the ISO-controlled grid within which congestion is expected to occur infrequently, so every bus will therefore have the same locational price. Four zones have been defined in California: Northern and Southern California, San Francisco, and Humboldt county. Only two of these zones have been active—Northern and Southern California—so the ISO has only calculated congestion charges between these two zones.

Though the ISO has ultimate responsibility for congestion management on the ISO-controlled grid, the PX and other scheduling coordinators have been the "market makers." One controversial aspect of separating these two responsibilities was the extent to which the ISO, an entity that is not supposed to be involved in commercial decisions, could use economic criteria to ration transmission resources. (In a nodal-pricing, pool-based system, the pool would have unlimited use of economic criteria to manage congestion.) Although the ISO in California is allowed to use some economic criteria (specifically, "adjustment bids" submitted by the scheduling coordinators from their specified "preferred schedule"), there has been concern raised that restrictions placed on the ISO might not result in least-cost congestion management (Stoft, 1996). The resulting dispatch and prices could be less efficient than if the ISO could adjust all resources economically.

In practice, to manage interzonal congestion in the day-ahead market, the ISO combines the preferred schedules from all scheduling coordinators to assess the feasibility of the combined schedule regarding coverage of losses, ancillary service requirements, reserve requirements, security criteria, and transmission capacity. When the aggregated schedule results in a congested interface between zones, the ISO uses "adjustment bids" to adjust schedules in the zones at the two ends of each path, and to determine the final day-ahead schedule and zonal prices to minimize total congestion costs. The zonal price differences, reflecting the use of adjustment bids, is then charged to all scheduling coordinators (including the PX, during its existence) as a transmission usage charge applied to all inter-zonal transmission flows.

In the hour-ahead market, if the combined schedules result in interzonal congestion, then the ISO will adjust schedules to relieve congestion. The resulting hour-ahead transmission usage charges are only applied to the difference between the interzonal flows in the day-ahead schedule and the actual real-time interzonal flow. Finally, if interzonal constraints appear in real time, the ISO can use its ancillary service generation or the final hour-ahead adjustment bids from scheduling coordinators to manage the constraint.

The main use of the zones is to determine the transmission usage charge across zones and to establish locational differentiation of power prices when interzonal congestion exists. The transmission usage charge is effectively a congestion charge collected by the ISO from the scheduling coordinators (including the PX). It is defined as the difference in zonal prices that is applied to the flow along the congested interties linking the congestion zones. Scheduling coordinators with schedules that relieve congestion on a congested interface receive a credit equivalent to the difference in zonal prices. Revenues collected from the transmission usage charge are credited against the revenue requirements of the various electric utilities and therefore reduce access charges.

Thus far, we have described the mechanics of interzonal congestion management and pricing. It is also possible, however, for congestion to occur *within* a zone. If congestion occurs within a particular zone (intrazonal congestion), the ISO uses adjustment bids to alleviate the congestion at minimum cost, and a zone-by-zone grid operations charge is imposed to collect the costs of using the adjustment bids from all transmission users based on their consumption.

In April 1999, the ISO approved the formation of firm transmission rights (FTRs; see Chapter 5, Section 5.3). An FTR is a contractual right that entitles the holder to receive a part of the usage charges collected by the ISO when interzonal congestion exists. FTRs allow the market participant that holds the interface rights to collect congestion charges whether or not it transmits power through that interface. The ISO conducts an annual FTR clearing price auction for each FTR market corresponding to different transmission paths from an originating zone to a receiving zone. FTRs allow market participants to hedge price risk associated with the incidence of congestion.

6.4.3. Transmission Losses

Each scheduling coordinator ensures that each generating unit for which it submits balanced schedules provides sufficient energy to meet both its demand and its estimated marginal contribution to transmission losses. Scheduling coordinators (1) can self-provide transmission losses by submitting a balanced schedule that includes the appropriate quantity of transmission losses, or (2) can settle obligations for transmission losses with the ISO using the real-time imbalance energy market.

Transmission loss responsibilities are determined through a power flow model that calculates a "generation meter multiplier" for each generator location. These multipliers can, in turn, be used to calculate the total demand that can be served by a given generating unit in a given hour, taking account of transmission losses.

6.4.4. Investment and Planning

Transmission planning and investment decisions for those utilities participating in the ISO were intended to be coordinated by the ISO, with participation from regional transmission planning organizations. The ISO, a participating utility, or any other market participant could propose to the ISO a transmission system addition or upgrade if it would promote economic efficiency or maintain system reliability. The ISO was expected, in cooperation with regional transmission organizations and state and federal regulators, (1) to determine when and where new transmission investment would be required and (2) to assign the costs to the various beneficiaries of the addition or upgrade in proportion to their net benefits. The utilities would then be required to make the necessary investments, the costs for which would be recovered from benefiting market participants and/or through the access charges. FERC can also require transmission investments. As a practical matter, however, few transmission investments have taken place since California's market opened for competition.

6.5. DISTRIBUTION NETWORK REGULATION AND RETAIL COMPETITION

Although industry restructuring in California required the IOUs to turn over transmission control functions to the ISO, the IOUs retained control over the distribution network, which remains regulated by the CPUC. Shorn of their transmission and generation control functions, these restructured "wires" utilities were called *utility distribution companies* (UDCs). In this section, we discuss the role and regulation of UDCs and the distribution services they provide, as well as their relationships with the competitive energy service providers (ESPs) that were allowed to compete for customers in California beginning in 1998.

6.5.1. Regulation of the Distribution Network

As noted earlier, the distinction between interstate and intrastate commerce in the United States forms the dividing line between federal (interstate) and state (intrastate) regulatory control. The distinction between interstate and intrastate electricity sales has, in general terms, been simplified into a distinction between wholesale and retail transactions. Whereas federal regulators have focused on wholesale transactions, state regulators in the United States have typically overseen utility distribution networks because these are more closely related to retail electricity consumption.

With competition, however, the precise dividing line between transmission and distribution systems has become more important. The FERC sought to clarify the distinction between federal and state authority in Order 888 by articulating seven criteria that distinguish local distribution facilities from interstate transmission facilities. These criteria form the basic dividing line between state and federal control in the United States and are the following.

1. Local distribution facilities are geographically close to retail customers.
2. Local distribution facilities are primarily radial in character.
3. Power flows into local distribution networks; it rarely flows out.
4. When power enters a local distribution system, it is not transported to some other market.
5. Power entering a local distribution system is consumed in a comparatively restricted geographical area.
6. Meters are based at the transmission/distribution interface to measure flows into the local distribution system.
7. Local distribution systems are reduced voltage systems.

The CPUC regulates distribution services based on these criteria. Under AB 1890, the UDC provides distribution services as a regulated monopoly and is responsible for maintaining the distribution system and responding to outages and other emergencies.

6.5.2. Remuneration for Regulated Distribution Activities

Historically, distribution revenue requirements have been established through ROR regulation and formal rate cases before the CPUC. Under this approach, the price of distribution service charged by a utility includes all of its variable and fixed costs plus a reasonable return on invested capital (see Chapter 4, Section 4.2).

More recently, however, the major electric utilities in California have been placed under Performance-Based Ratemaking (PBR), which decouples utility profits from costs and, instead, ties profits to performance incentives. This decoupling is accomplished by decreasing the frequency of rate cases, employing external measures of cost to set rates, or a combination of the two (see Chapter 4, Section 4.3, and Comnes, Stoft, Greene, and Hill, 1995). These systems are intended (1) to reward performance and therefore result in greater productivity and lower costs over time and (2) to reduce the frequency of complex and costly ratemaking procedures. In practice, there are different types of performance-based regulation, including price caps, revenue caps, sliding scale, and targeted incentive ratemaking (see Chapter 4, Section 4.3).

Although the use of and benefit from PBR was articulated and reaffirmed throughout the state's restructuring process (beginning in 1994), PG&E, SCE, and SDG&E had filed initial PBR plans in 1992 and 1993 (Che and Rothwell, 1995; EIA, 1998). The PBR plans ultimately adopted by the CPUC for each of these three utilities differ, sometimes significantly (SDG&E has a revenue cap, whereas SCE and PG&E use price caps), and the distribution PBR mechanisms are still being refined. Despite differences, a CPUC study points out that the plans all include (CPUC, 1997c):

- Formulas to establish revenue requirements or rates that are indexed to inflation and adjusted for productivity changes and changes in the cost of capital
- A revenue-sharing mechanism allowing shareholders and ratepayers to share actual revenues compared to those authorized
- A reward and penalty system to ensure that employee safety, reliability, and customer satisfaction standards are maintained compared to established benchmarks
- Inclusion of adjustments ("Z" factors), to capture the influence of exogenous factors not under the utility's control
- A monitoring and evaluation program

Another aspect of remuneration for distribution activities stems from the potential for distribution system line losses (and losses associated with meter error and energy theft). To account for these energy losses, customer-class-specific "distribution loss factors" are calculated by the three major IOUs, and are used for scheduling and settlement purposes (Distribution Loss Factors Working Group, 1998).

6.5.3. Retail Competition

As noted earlier, AB 1890 and related CPUC rulings gave all retail customers in California's IOU service territories the opportunity to choose to have their electricity and their meter and billing services provided by an independent ESP beginning in 1998. (As discussed below, this option was eliminated in 2001 in response to the electricity crisis.) But Californians were not compelled to switch power providers, and could opt to continue to have their power and meter/billing needs met by their UDC, subject to the legislatively mandated 10% rate reduction discussed in Section 6.2.2.2 (also see Section 6.6.1, below). Regardless of who supplies customers with their electricity, however, all customers were required to pay their local UDC's transmission and distribution fees and other nonbypassable fees, e.g., the CTC.

Those customers who did choose to switch power providers could conduct business with one of several unregulated ESPs. These entities are unregulated inasmuch as they are not subject to the same regulation as the UDCs. But ESPs were required to meet specific criteria to participate in the market and their behavior was subject to review by the CPUC, the Federal Trade Commission (FTC), and other consumer protection bodies. Below we review the role of ESPs and related entities offering both energy and meter/billing services.

6.5.3.1. ESPs and Competitively Offered Energy Services. To offer services in the retail market, ESPs were first required to enter an UDC–ESP service agreement that, at a minimum, includes

- ESP identification and contact information
- A warranty that the ESP has obtained a certified scheduling coordinator
- An agreement on the provision of meter and billing services including data protocols for the exchange of meter and billing data

In addition, if the ESP was seeking to serve residential or small commercial customers (defined as customers with less than a 20 kW demand) AB 1890 required that the ESP (1) register with the CPUC (providing contact information, legal details and history, and evidence of firm power supply contracts or sources), and (2) employ an authorized *independent verification agent* (IVA) who would contact and independently verify the decision of customers under 20kW seeking to switch suppliers. The purpose of the IVA was to protect small customers from unscrupulous or high-pressure sales tactics.

Once the ESP had taken these steps, it could begin submitting *Direct Access Service Requests* (DASRs) to the UDC. ESPs were required to submit a DASR for each new customer with direct access service. DASRs generally had to include customer name, service account address, UDC service account number, ESP name, ESP registration number (if applicable), metering service option and equipment needs (if applicable), meter identification information (if not a UDC meter), and billing service option. Once the DASR had been processed, the customer account was switched and the customer began taking service from the new ESP. In the event that the ESP was unable to meet its energy or financial commitments, the ESP's customers would default to the UDC.

6.5.3.2. Billing and Metering Services. The CPUC determined in 1997 that ESPs and related entities should be free to compete with UDCs in offering meter and billing services as well as power supply (CPUC, 1997a; CPUC, 1997b). Under the CPUC's rulings on meter and billing services, ESPs could offer three billing options:

1. Consolidated UDC billing, in which the UDC bills for both ESP and UDC services
2. Separate billing, in which the ESP and UDC send separate bills for their respective services
3. Consolidated ESP billing, in which both ESP and UDC charges are billed by the ESP

The CPUC also determined that ESPs could provide metering services subject to specific requirements. The ESPs and UDCs providing metering services were ultimately responsible for collecting, transferring, and processing meter data. However, the provision of meters and the collection of meter data was done by *Meter Service Providers* (MSPs), who had to be certified, and *Meter Data Management Agents* (MDMAs), who were screened by the UDCs. Typically, these entities are specialized companies whose services are contracted for by the UDCs or ESPs. Although competition was allowed in the provision of meters and in meter reading technology, all metering entities had to abide by open architecture standards at specific points in the meter data flow network (CPUC, 1997d). These open architecture standards were designed to ensure that incompatible and/or entirely proprietary metering systems would not develop.

6.6. PARTICULAR ASPECTS OF THE REGULATORY PROCESS IN CALIFORNIA

6.6.1. Stranded Costs

One of the most contentious parts of the electricity restructuring process in California was the recovery of "stranded costs." Under historic ROR regulation, the CPUC allowed utilities to collect revenues for those costs prudently incurred to serve customers, including a reasonable profit on and repayment of their capital investments. In a competitive market, the market sets prices, hence revenues. These revenues might not be sufficient to provide utility shareholders with a return on their original investment. Roughly speaking, then, "stranded costs" are simply the difference between regulated retail electricity prices for generation services and the competitive market price of power.

The level of total stranded costs in California depended on the market price of electricity. At the onset of electricity restructuring, when PX prices were expected to average approximately $24/MWh, SCE, PG&E, and SDG&E estimated their total stranded costs to be $26.4 billion. With higher PX prices experienced during the early years of market operations, however, the utilities' total stranded costs were much lower.

Because California utilities' past investments were made as part of the previous "regulatory compact" and were approved by the CPUC, the CPUC and the state legislature determined that it would be unfair to penalize utility shareholders and bondholders for these past investments. Stranded cost recovery was therefore allowed, paid by *all* customers through a separate per-kWh charge on electricity bills. As noted earlier, this charge was the ***Competition Transition Charge*** (CTC), and was to be levied on customers from 1998 through 2001. The charge was nonbypassable, and could not therefore be avoided by switching electricity providers.

The utilities were not, however, strictly guaranteed to recover all of their stranded costs. Retail electricity rates were originally fixed for the duration of the stranded cost recovery period (1998–2001). Revenue available for stranded cost recovery was therefore equal to the difference between the retail electricity rates to final customers (capped at the 1996 level minus the 10% reduction) and the cost of meeting the demand (the energy costs plus transmission and distribution costs). If the difference between these two quantities was not sufficient to pay off all stranded costs, the utilities were required to write off the remaining amount. If, on the other hand, the utilities could collect all of their stranded costs before 2001, the rate freeze was to end. SDG&E, for example, completed its stranded cost recovery in 2000, ending its rate freeze.

6.6.2. Market Power

The goal of electricity industry restructuring is to move from a regulated utility monopoly structure to a "workably competitive" marketplace. But restructuring will not be in the public interest if it allows some companies to exploit market dominance and to stifle competitive market forces. Market power is the ability of one firm, or a set of firms, to unduly influence prices, quantities, product quality, and other conditions in a particular market (see Chapter 2). In the past, because of extensive state and federal regulation, market power was not considered a significant problem. At least three types of market power can distort competition in electric power markets:

1. **Vertical market power**, resulting from ownership or control by a single firm of more than one phase of electricity production (generation, transmission, distribution). Vertical integration and market power could allow a firm to erect barriers to entry or otherwise shift costs and revenues among affiliates in ways that distort efficient market operation.
2. **Horizontal market power**, resulting from concentration of ownership or control of any single phase of electricity production, such as generation. For example, it would allow generators to withhold generation or bid strategically to force higher market-clearing prices.
3. **Locational market power**, created when a specific generation facility provides unique services (e.g., reliability) needed for a particular geographic region and thus can command a premium above the market price.

California's restructuring legislation and regulations imposed many requirements to reduce the potential for the exploitation of market power. For example, California's legislative and regulatory bodies

- Created an ISO to operate the utilities' transmission systems and to control vertical and locational market power
- Required the mitigation of locational market power by requiring generators that are needed to solve local reliability problems to enter into RMR contracts with the ISO, thereby fixing the amount that a generator is paid when operated for reliability reasons (for a detailed discussion of the functioning and problems with this market, see ISO, 1999a)
- Called for the utilities to divest of much of their generating assets to reduce concentration in generation and, therefore, horizontal market power
- Required the IOUs temporarily to bid all of their generation into the PX and to purchase all of their electricity from the PX to further mitigate horizontal market power and self-dealing
- Called for the functional unbundling of generation, transmission, and distribution
- Established affiliate rules that do not allow affiliated, unregulated companies of the regulated utilities to unduly use their market position to restrict competition
- Established price caps on various ISO markets

As discussed in Section 6.7, these measures were ultimately insufficient in mitigating the exercise of market power in the California market.

6.6.3. Public Purpose Programs

AB 1890 established public funding mechanisms to continue the support of a range of activities that fall under the loose heading of public purpose programs. These programs include activities designed to support or promote (1) low-income customer assistance, (2) energy efficiency, (3) renewable energy, and (4) research, development, and demonstration (RD&D) activities.

These programs were historically mandated by the CPUC and the state legislature and administered by the utilities themselves, subject to regulatory oversight. This arrangement allowed the costs incurred by utilities in facilitating or providing public-purpose programs to be recovered as part of their regulated monopoly rate base.

Concerned that the important public benefits these programs provide would be lost in the new market structure, the legislature mandated that funding for public purpose programs come from a nonbypassable charge on consumers' bills. AB 1890 specifies that monies be collected and used as follows:

- Low-income customer assistance was to continue indefinitely, funded at no less than 1996 levels (roughly $324 million).
- Energy-efficiency programs were to be supported from a fund of $872 million collected from 1998 to 2001.

- Renewable energy technologies were to receive assistance from a fund of $540 million collected from 1998 to 2001.
- Research, development, and demonstration programs were to be supported from a fund of $250 million collected from 1998 to 2001.

AB 1890 gave the CPUC control of the low income and energy efficiency funds and gave the CEC control over the renewable energy and RD&D funds. Illustrating the importance of these public purpose programs, legislation was passed in 2000 to extend funding for these programs through 2012 at annual funding levels similar to those established for the initial 4-year transition period.

6.6.4. Customer Protection and Small Customer Interests

Finally, under AB 1890 it was deemed important to address potential equity and distributional implications of electricity industry restructuring. At the onset of reform in California, many believed that, due to their relative lack of bargaining power, small customers (especially residential and rural customers) would not see as many benefits from restructuring as larger customers.

In California, the legislature wanted to ensure at least some benefits for smaller customers. Consequently a mandated initial 10% rate reduction from 1996 rates took place in January 1998 for residential and small commercial (i.e., less than 20 kW) customers. Based on the legislation, rates were to be reduced by a further 10% in 2001. As noted earlier, to help finance the rate reduction, AB 1890 authorized the IOUs to issue *rate reduction bonds*, the proceeds of which would be used to help pay a part of the utilities' stranded costs. Because the bonds carry a lower interest rate and a longer term than otherwise would have been feasible, they allowed an immediate 10% rate reduction.

In addition to a guaranteed rate cut, AB 1890 also mandated supplier price and fuel source disclosure requirements, registration requirements, and the continuation of the universal service requirement. The CPUC also put in place affiliate transaction standards. These standards were designed to prevent the incumbent utility from abusing its position as the distribution company to encourage customers to take service from its unregulated subsidiary.

6.7. MARKET EXPERIENCE AND THE ENERGY CRISIS

This section presents early market operations experience with California's restructured electricity sector, and highlights the nature and causes of and solutions to the state's energy crisis. Because California's electricity sector is in such flux, we keep this section brief and simply summarize the key points. We refer readers to other documents for more in-depth and updated analyses of experience with California electricity reform and the resulting crisis (see, e.g., Joskow and Kahn, 2000; Besant-Jones and Tenenbaum, 2001; Borenstein, 2001; Joskow, 2001; Marcus and Hamrin, 2001; McCullough, 2001; and Sweeney, 2002).

6.7.1. Market Operations: 1998 and 1999

During the first two years after market reform in 1998, there was limited evidence of the problems that would consume California's electricity sector beginning in the summer of 2000. But even during these first two years of market operations, some problems were evident. First, electric generators were apparently able to raise prices above competitive levels (Borenstein, Bushnell, and Wolak, 2000). Second, the ISO was constantly redesigning the ancillary services markets in the face of bid insufficiency and considerable price volatility, prompting the ISO to institute price caps (Siddiqui, Marnay, and Khavkin, 2000; ISO, 1999b). Finally, customer switching to competitive ESPs remained low during these early years of reform. At its peak, about 10% of eligible load had switched providers in California: 16% of eligible industrial customers switched providers, for example, as did approximately 2% of residential customers. (For reviews of ESP product offerings, customer switching, and customer switching experience, see Wiser, Golove, and Pickle, 1998, and Golove, Prudencio, Wiser, and Goldman, 2000).

Despite these concerns, however, reliability remained high and market prices were low. From August 1, 1998, to June 30, 2000, 99% of the energy traded in the PX was in the day-ahead market. During the same period, the total amount of energy traded in the PX was equal to 370 TWh, with prices averaging $33.84/MWh and $37.63/MWh in the day-ahead and hour-ahead/day-of markets, respectively. The PX day-ahead and hour-ahead/day-of unconstrained markets also performed with predictable, seasonal, and daily patterns. Prices rose considerably in summer months and decreased during the winter.

6.7.2. The Electricity Crisis

Beginning in the summer of 2000 and continuing through the first half of 2001, California's electricity system was in crisis. The full effects of this crisis were seen in many ways:

- **Wholesale Power Prices Skyrocket.** Beginning in June 2000 and continuing through the first half of 2001, wholesale electricity prices rose to unprecedented levels. By way of example, the total cost of wholesale electricity procurement to meet load in the ISO's control area during 1999 was $7.4 billion. In 2000, that cost rose nearly fourfold, to $28 billion. Wholesale electricity prices, as seen in the PX day-ahead market and the ISO's real-time market, sustained price levels that exceeded $100/MWh and in some months average prices exceeded $300/MWh.
- **Electricity Reliability Degrades.** Simultaneously, electricity reliability got progressively worse. Stage 3 emergencies are the highest level of electrical emergency called by the ISO, and are triggered when operating reserves fall below 1.5%. The ISO called no such emergencies in 1999. One Stage 3 emergency was called during 2000. During the first quarter of 2001, the ISO called 36 such emergencies, triggering rolling blackouts several times to match supply and demand in the state.

- **Financial Catastrophe for Utilities.** Required by AB1890 to purchase power from the PX spot market to meet their electricity needs, and to sell that power at capped retail electricity rates, the in-state electric utilities were deeply in debt by the end of summer 2000, because wholesale procurement costs far outstripped retail electricity revenues. By 2001, in-state utilities had incurred billions of dollars of debt, making them unable to secure credit for further power purchases. In early 2001, the state government began to purchase electricity for these utilities. On April 6, 2001, PG&E filed for bankruptcy.
- **Retail Access Dies.** With massive regulatory uncertainties and unprecedented wholesale electricity prices, competitive electricity service providers largely withdrew from the state in late 2000 and 2001, turning customers back to the utilities for default service. Legislation and regulation in 2001 suspended retail access, ending California's experiment with retail choice.

6.7.3. The Causes of the Electricity Crisis

The causes of California's electricity crisis are several. There is no consensus on the relative importance of various causal factors. The more important causes typically implicated in the crisis are highlighted here.

6.7.3.1. Market Fundamentals

- **Supply–Demand Balance.** Demand growth throughout the Western United States outstripped new plant construction during the 1990s, resulting in a tightening of the supply–demand balance. Average demand growth in California from 1996–2000 was 2.5% per year. Importantly, region-wide demand growth and the lack of hydroelectric supply from the Pacific Northwest resulted in deep cuts in the electricity imports that California historically relied upon during the summer months.
- **Natural Gas Prices.** Natural gas prices, which normally range from $2 to $3/MBtu, increased nationwide but especially in California, where pipeline capacity constraints are significant and market power abuses have been claimed. Natural gas prices peaked at $60/MBtu in late 2000 and were consistently above the national average from mid-2000 to mid-2001.
- **Emissions Credits.** Power plants in the Los Angeles basin are required to purchase emissions credits to offset their own pollutant emissions. The cost of these emissions credits increased dramatically in the year 2000 with the increased use of in-basin power plants.

6.7.3.2. Market Structure

- **PX Buy–Sell Requirement.** Electric utilities, required initially to purchase their power from the PX spot market, while selling their power at capped retail electricity rates, were largely unable to enter long-term hedging contracts to mitigate price volatility and protect against high spot-market prices.

- **Generation Divestiture without Buy-Back Contracts.** Utilities divested a significant amount of their generation capacity, without long-term buy-back contracts. This exacerbated the exploitation of market power by generators and offered no price stability or certainty for utility electricity purchases.
- **Retail Rate Freeze and Little Demand Response.** A competitive market generally requires responsive supply and demand. With a retail electricity rate freeze, and with the market rules established at the ISO, little economic opportunity for demand responsiveness existed.
- **Uncoordinated Maintenance Schedules.** Uncoordinated generator maintenance schedules helped contribute to a lack of supply during winter months, when demand is generally low in the state, as multiple generation units were down for maintenance simultaneously.

6.7.3.3. Market Power. Concentration in generation plant ownership, a large amount of unhedged power purchases, limited demand response, and tight supply conditions offered electricity generators the ability to exert market power. The withholding of generation capacity, either physically or economically, was of particular concern, and many studies have detected exploitation of market power (e.g., by Enron). Market power exploitation appears to have been a significant factor in raising wholesale power prices (see Borenstein, Bushnell, and Wolak, 2000).

6.7.3.4. Regulatory and Political Inaction

- **Policing Market Power.** Analysts have pointed to FERC's apparent inability or unwillingness to police market power abuses, through price caps or other measures, as a contributing factor to the duration of the crisis. FERC did not implement these price caps until June 2001.
- **Long-Term Hedging Contracts.** The CPUC and California State Legislature's inaction in quickly allowing and approving the use of long-term, power-purchase hedging contracts contributed to the crisis.
- **Rate Freeze and Demand Response.** Finally, an initial unwillingness by state policymakers to end the rate freeze and raise retail electricity rates to reflect costs, thereby also stimulating demand response, deepened the financial crisis for the utilities and did nothing to ease the supply–demand imbalance.

6.7.4. Solutions and Conclusions

By mid-2001, the electricity crisis in California had eased. Wholesale electricity rates declined and expected rolling blackouts during the summer of 2001 did not happen. A combination of factors helped ease the crisis. These included

- Substantial reductions in natural gas prices
- Lower electricity demand than expected, in part due to conservation, energy efficiency, and load management measures (encouraged by substantial in-

creases in the budgets for energy efficiency and education programs, as well as by increases in retail electricity rates)

- An increase in the supply of power, in part due to new plant construction
- The FERC's setting of wholesale electricity caps

In addition to these factors, and because of the weakened finances of the electric utilities, a state agency (Department of Water Resources, DWR) stepped in to purchase wholesale electricity for the utilities. This agency signed many short and long-term contracts with electricity generators to meet demand projections and reduce the ongoing role of volatile spot-energy transactions. The cost of these contracts, many of which were priced above historic market levels, will be passed on to California electricity consumers.

As the utilities regain financial strength, they will again regain their historic role of procuring electricity for their customers and the role of the state government will decrease. What remains unclear is how these procurements will be designed, how the wholesale electricity market will be structured and operated, and whether retail electricity competition will be reintroduced.

California's electricity crisis resulted from a combination of factors. Regardless of the causes and ultimate resolution, this crisis has had substantial implications for the design of future competitive electricity markets. Market designers must understand the California case before restructuring regulated electricity industries elsewhere. We hope this case study provides a foundation for this understanding.

CHAPTER 7

THE NORWEGIAN AND NORDIC POWER SECTORS

Helle Grønli

Norway was among the first countries in the world to open its electricity industry to competition. The principles of the Norwegian market design have been implemented in neighboring Scandinavian countries. Scandinavia now has the world's first international power pool. As opposed to England and Wales, which deregulated in 1989, the basic principles of the Norwegian market design have remained unchanged since 1991. In 2000, England and Wales changed their market organization to a model similar to the Scandinavian model.

7.1. GENERAL DESCRIPTION OF THE NORWEGIAN POWER SYSTEM

7.1.1. Generation

Nearly 100% of Norway's generation is based on hydroelectric energy (OED, 2000). The remainder is based on thermal power (combustion technology). The total installed generation capacity for hydroelectricity was 27,470 MW as of January 1, 2000, whereas the installed capacity for thermal power was 293 MW. In addition, there was 13 MW of wind power generation. Total electricity generation can vary extensively from year to year due to the high dependency on hydroelectricity. Generation averages approximately 118 TWh annually, but can vary from approximately 98 to 148 TWh in dry and wet years, respectively. Reservoirs associated with power generation had a total capacity of approximately 84 TWh in early 1999.

Norwegian energy policy implies increased generation from renewables, targeting 3 TWh from wind power annually by year 2010. There are also ongoing discussions regarding natural gas power generation. Norway is a large gas producer in Europe, and several market participants have shown interest in establishing gas power facilities recently. The regulator, the Norwegian Water Resources and Energy Directorate (NVE), issued concessions allowing gas power stations with a total installed capacity of 770 MW in 1997. More applications are being investigated. However, the profitability of these planned facilities is unclear.

There are 158 generators in Norway, of which Statkraft SF (a state-owned company) is the largest, with an average annual production of 34 TWh. Of these generators, 27 are producing electricity for industrial purposes to cover their own consumption. Table 7.1 summarizes the mean annual production for the 10 largest generators in Norway. Other generators (*wholesale companies*) own and operate generation and regional grids. These generators usually sell their generation to local distribution companies, without dealing with end-users. There are 22 of these wholesale companies in Norway (OED, 2000).

7.1.2. Transmission

Statnett SF, the state-owned national grid company, operates the transmission grid in Norway. Statnett owns approximately 80% of the transmission grid that it operates, whereas regional grid owners own the remainder. Through the *Central Grid Agreement*, Statnett rents additional transmission capacity from regional grid owners. The Norwegian transmission grid consists of 420 kV, 300 kV, and parts of the 132 kV grid. Furthermore, Statnett is the Norwegian Transmission SO, being responsible for system dispatch, ancillary services, etc. Table 7.2 shows transmission facilities in Norway and generation connected to the National Central Grid.

Table 7.3 shows existing and planned interconnections between Scandinavia and Continental Europe. The total capacity between Sweden and Finland is 1200 MW to Sweden and 2000 MW to Finland. The maximum capacity between Norway and Sweden is approximately 4000 MW. This capacity is, however, influenced by total system flow and possible congestion at other interfaces in the system, i.e., transfer between Norway and Sweden can never exceed 3000 MW. Denmark, however, con-

Table 7.1. The 10 Largest Generators in Norway as of January 1, 1999

	Mean Annual Generation [GWh]	Percent of Total Generated Energy	Installed Capacity [MW]	Percent of Total Installed Capacity
Statkraft SF	33,828	30.4	8,736	32.0
Oslo Energi Produksjon AS	6,912	6.2	2,098	7.7
BKK Produksjon AS	5,911	5.3	1,500	5.5
Lyse Energi AS	5,061	4.5	1,484	5.4
Norsk Hydro Produksjon AS	4,479	4.0	804	2.9
Trondheim Energiverk AS	2,922	2.6	725	2.7
Hafslund Energi ASA	2,653	2.4	545	2.0
Vest-Agder Energiverk	2,547	2.3	614	2.2
Kraftlaget Opplandskraft	2,462	2.2	522	1.9
SKK AS	2,432	2.2	581	2.1
Total for the 10 largest	69,117	62.1	17,609	64.4

Source: OED (Department of Energy and Industry, 2000).

Table 7.2. Transmission Facilities in Norway as of January 1, 1998

Grid level	420 kV	300 kV	132 kV
Transmission grid	2,125 km	5,399 km	3,627 km
Regional grid	0	0	5,387 km
Connected generation	20%	34%	34%

Source: Norwegian Water Resources and Energy: www.nve.no

sists of two systems: Western Denmark is connected to the UCTE (Union for the Coordination of the Transmission of Electricity) system, whereas Eastern Denmark is connected to the NORDEL (Nordic Organization for Electric Cooperation) system.

In Table 7.3, capacities in parentheses illustrate interconnections being discussed. These interconnections are the Storebælt (Denmark–Denmark), the Euro Cable (Norway–Germany), and the Viking Cable (Norway–Germany). Besides the planned interfaces included in Table 7.3, the NorNed cable will connect Norway and the Netherlands at a minimum of 600 MW, the North Sea Interconnector will connect Norway and England, and the SwePol Link will connect Sweden and Poland with 600 MW.

Electricity exchanged with neighboring countries varies a great deal from year to year due to precipitation. This can be seen in Table 7.4. It shows the net exchange between Norway and neighboring countries.

7.1.3. Distribution

The Norwegian power system has two grid levels besides the transmission grid:

1. Regional grids (60 – 132 kV)
2. Distribution/local networks (≥ 22 kV)

Table 7.3. Existing and Planned Interconnections in Scandinavia, 1999 [MW]

	To:					
From:	Norway	Sweden	Finland	Denmark	Germany	Russia
Norway	—	4,200	70	1,040	(min 600)	50
Sweden	4,000	—	2,085 (80)	2,680	600	
Finland	70	1,485 (80)	—	—		60 (300)
Denmark	1,040	2,640	—	(500-600) West-East	1,800	
Germany	(min 600)	600		1,400	—	
Russia	50		1,160 (300)			—

Source: NORDEL (2001).

Table 7.4. Net Exports (–) and Net Imports (+) from 1991–1999 [GWh]

Year	Sweden	Denmark	Russia	Finland	Total
1991	–1,832	–1,018	9	–79	–2,910
1992	–5,709	–3,047	31	–99	–8,824
1993	–5,810	–1,952	0	–26	–7,788
1994	–1,561	1,194	0	292	–75
1995	–6,143	–1,017	80	11	–7,069
1996	3,939	4,680	176	252	9,047
1997	3,148	639	50	180	4,017
1998	4,375	–909	193	19	3,678
1999	25	–2,137	232	–3	1,883

Source: www.statnett.no and NORDEL (2001).

The regional grid companies are normally vertically integrated with generation. These are the *wholesale companies.* There were 190 distribution network companies in 1999, of which approximately 50% are integrated with generation and/or a regional grid. Table 7.5 shows total installations in the distribution network in early 1999. Table 7.6 shows the 10 largest distribution companies.

7.1.4. Consumption

Electricity-intensive industry consumes 28% of total consumption in Norway, whereas residential customers use another 36%. Commercial customers consumed the remainder of a total annual consumption of 120,600 GWh in 1998. Table 7.7 shows the total consumption for the period 1992–1998. The 190 distribution network companies served 2,487,888 customers in 1998 (NVE, 2001).

7.1.5. Economic Indices

The electricity industry made up 1.9% of the Norwegian GDP in 1999, which was approximately 23 billion Norwegian Kroner (NOK, or US$2.7B at an exchange rate

Table 7.5. Distribution Network Owners' Installations as of January 1, 1999

Distribution Installation		Total
Lines	High-voltage	63,867 km
	Low-voltage	115,759 km
Underground cables	High-voltage	26,821 km
	Low-voltage	70,004 km
Sea cables	High-voltage	1,546 km
	Low-voltage	186 km

Source: Norwegian Water Resources and Energy: www.nve.no

Table 7.6. The 10 Largest Distribution Companies as of December 31, 1998

Distributor	Number of Customers	Total Sales (GWh)
Viken Energinett AS	303,726	8,552
BKK Distribusjon AS	128,719	3,349
Østfold Energi Nett AS	88,642	2,144
Trondheim Energiverk Nett AS	83,119	2,216
Nord-Trøndelag Elektrisitetsverk	74,412	1,991
EAB Nett AS	72,183	2,282
Troms Kraftforsyning	60,439	1,947
Vest-Agder Energi Nett AS	57,402	1,353
Stavanger Energi AS	55,825	1,983
SKK Nett AS	53,304	1,449

Source: OED (2000).

of U.S.$1 = 8.5 NOK). Total investments in the electricity sector were 4.5 billion NOK (US$534M) in 1999. However, investments in the electricity industry have been declining.

Prices can vary extensively between years and seasons in Norway. Weekly average prices over the period 1992–2001 are illustrated in Figure 7.1. Temperature and precipitation are the main influences on electricity prices in Norway. For instance, the price spike during the Lillehammer Olympics in 1994 was caused primarily by very cold weather. Low precipitation can explain the high prices of 1996. Compared with the seasonal and annual variations, daily variations are small in Norway. Figure 7.2 illustrates hourly prices from three arbitrarily chosen days of 1999.

Table 7.7. Gross Annual Consumption 1992–1998 [GWh]

Year	Total Consumption
1989	104,300
1990	106,000
1991	108,200
1992	108,900
1993	112,100
1994	113,200
1995	116,400
1996	113,800
1997	115,500
1998	120,300

Source: Norwegian Water Resources and Energy: www.nve.no. *Note:* Grid losses are included.

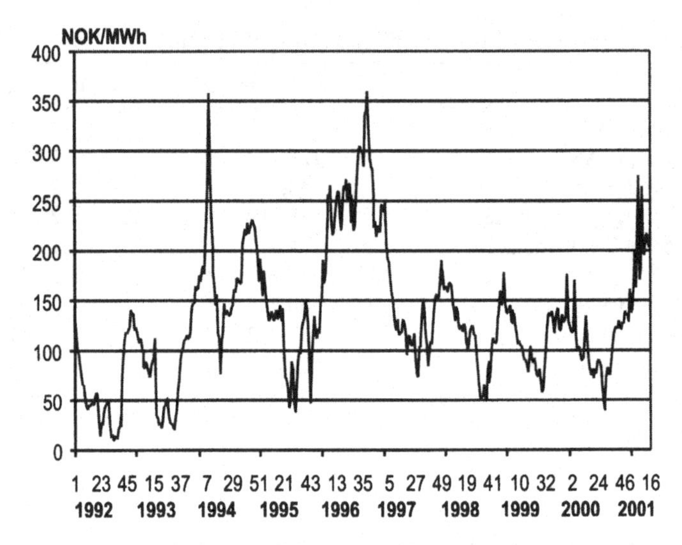

Figure 7.1. Weekly average spot prices, 1992–2001. (Source: Nord Pool.)

7.1.6. General Economic and Energy Indices (*Sources: SSB (2002) and EIA: www.eia.doe.gov/emen/cabs*)

- Population (2002): 4,524,066
- Size: 0.4 million square kilometers with Jan Mayen and Svalbard
- Gross Domestic Product (2000E): 1347 billion NOK (U.S.$151.9B)
- Real GDP growth rate (2000E): 3.3%

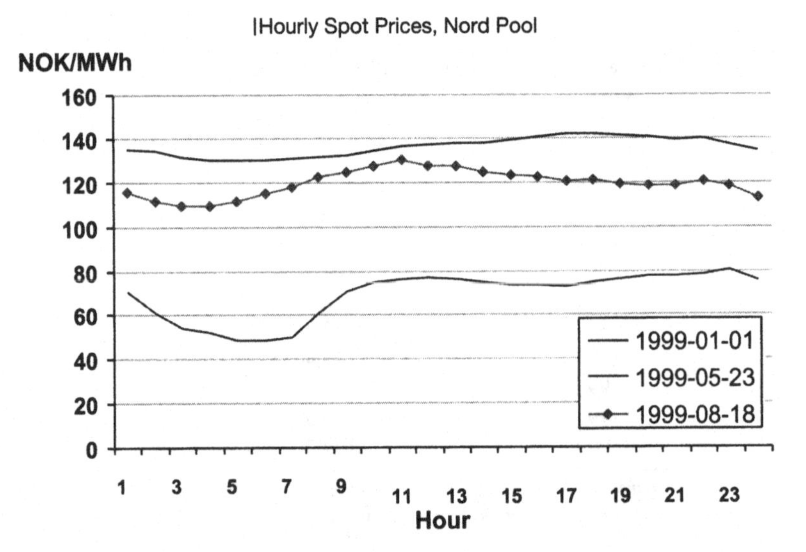

Figure 7.2. Hourly spot prices three days of 1999. (Source: Nord Pool.)

- Inflation rate (2001E): 3.0%
- Oil production (2001): 3.4 million barrels per day (bbl/day)
- Net oil exports (2001): 2.9 million bbl/day
- Natural gas production (2001): 1.61 trillion cubic feet (Tcf)
- Natural gas consumption (2001): 0.01 Tcf
- Net natural gas exports (2001): 1.4 Tcf
- Coal production (2001E): 1.73 million short tons
- Coal consumption (2001E): 1.71 million short tons
- Electricity production (2001): 122 Twh
- Electricity consumption (2001): 113 Twh (without grid losses)

7.2. THE NEW REGULATORY FRAMEWORK

The Norwegian electricity industry has traditionally been publicly owned (state, county, and municipality). This decentralized organizational structure was maintained until the Energy Act of 1990 became effective in January 1991. The reform required no change of ownership, although the functions of grid operations were separated from competitive functions through separate financial reporting. NVE was appointed regulator upon deregulation. The Norwegian Competition Authority was given control of competition in wholesale and retail markets. A description of the legal and regulatory framework is given in Table 7.8.

The first power plants in Norway were built in the late 19th century, and electrification first took place in the cities. Utilities were municipally owned and aimed at

Table 7.8. The Regulatory Framework in Norway

Regulatory Framework	Responsible Institution	Description
The Energy Act of 1990	The Parliament	The overall framework for the deregulated electricity market
Governmental Regulations on: General issues detailing the Energy Act Regulation of grid companies, financial and technical reporting and calculation of tariffs; Metering and accounting	The Government	Specifies aspects of the Energy Act
Guidelines on: System operations and Energy efficiency	The Regulatory Authorities (NVE)	NVE is given the authority to develop guidelines within the framework of the Energy Act.
Concessions	The Regulatory Authorities (NVE)	NVE is given the authority to give concessions for distribution, trade, generation, and system operations.

offering electricity as cheaply as possible. Between 1887 and 1894, the Norwegian Parliament passed legislation for the expropriation of dams and watercourses for hydroelectric generation, as well as sites for building transmission facilities. Early in the 20th century, rural areas were electrified and more large-scale technology used. Still, due to the nationwide distribution of hydroelectric resources, both large- and small-scale hydroelectric stations were built.

Private exploitation of the hydroelectric resources was strictly regulated as a response to increasing foreign investment in hydroelectric generation and power-intensive industry (such as the electrochemical and electrometallurgical industries). The power-intensive industries generated much of their own electricity in the early years. However, since World War II, they have received inexpensive electricity from the Norwegian State Power Board (now known as Statkraft) through politically determined prices. Although local initiative was the main driver behind establishing the electricity industry, municipal utilities started cooperating as early as 1932, when the first of five regional, but not interconnected, power pool organizations was established to solve common tasks. A joint power pool for the whole country, Samkjøringen, replaced these five regional power pools in 1971.

7.2.1. The Energy Act of 1990: Objectives and Consequences

The Energy Act of 1990, which introduced deregulation in the Norwegian electricity industry, became law on January 1, 1991. The basic idea behind the reform was to unbundle functions and processes to expose some to competition, without privatizing the industry. Natural monopolies remained regulated, and NVE was appointed the regulator of transmission and distribution services. The intentions of the Energy Act are summarized below.

- **Avoid excessive investment.** Prederegulation, utilities could automatically pass through the costs of investments to their customers on a cost-recovery basis, without being measured against competitors and without facing the risk that investment decisions can bring. Avoiding excessive investments should increase efficiency in the market, and eventually result in lower prices to end-users.
- **Improve selection of investment projects by choosing the most profitable ones first.** The combination of municipal owner structure, political constraints, and an obligation to serve customers within each local area resulted in the suboptimization of hydroelectric investment projects. The prederegulation, distance-dependent tariffs added to the tendency toward local thinking rather than national thinking about generation expansion.
- **Create incentives for cost reductions through competition.**
- **Fair cost distribution among customers, i.e., avoid cross-subsidization between customer groups.** Electricity prices have traditionally been determined through political processes in Norway. A result of this has been cross-subsidization among customer groups. From a macroeconomic perspective, this leads to a loss of social welfare.

- **Reasonable geographical variations in prices.** Prices have traditionally varied extensively between local areas due to Norway's diverse geography and varying conditions for hydroelectric generation and transmission. Customers, therefore, received unclear and opposing signals about the value of electricity.

Some consequences of the Energy Act of 1990 have been the following:

- The high-voltage transmission system of the state-owned power company, Statkraft, was divested into a separate company, Statnett. Statkraft was reorganized to become a commercially oriented generator. Statnett was given the responsibility of operating the transmission grid formerly owned by Statkraft. Both companies have remained state owned.
- Third-party access was introduced at all grid levels. All grid owners have an obligation to connect to the transmission grid and serve customers in the local area. No discrimination against deliveries from any supplier is allowed.
- Retail access was introduced for all customers, including residential customers.
- A Norwegian power pool, known as Statnett Marked, was designed as a Statnett subsidiary. (It was later renamed Nord Pool.)
- Vertically integrated companies were required to separate financial accounting for competitive and regulated functions.

The Energy Act provides the legal foundation for promoting an efficient electricity market and encouraging flexible energy use. The Act does not detail how to implement this. Three important governmental regulations supplement the Energy Act. When deregulating, NVE was given the authority of issuing guidelines and bylaws to support the intentions of the Energy Act. NVE regulates the functions remaining under monopoly control. In addition, NVE and the Norwegian Competition Authority supervise competitive functions. NVE issues licenses for trade, generation, transmission, distribution, and system operations requiring adherence to specific rules. The licensees have the opportunity to appeal NVE's decisions to the Department of Energy and Industry. The Government and NVE have developed a set of regulations and guidelines for the industry, as described in the following sections.

7.2.2. The Energy Act of 1990: Specifics

7.2.2.1. Governmental Regulations Detailing the Energy Act. The governmental regulations (NVE, 1999) define in more detail what is required from licensed companies, specifying, among other things, the intentions of the Energy Act regarding customer protection. The requirement of financial separation between the regulated and competitive functions is found in these regulations. The regulations also outline the main principles of monopoly regulation, including defining the grid companies' minimum allowed profit.

7.2.2.2. Governmental Regulations on Monopoly Regulation and Tariffs. The governmental regulations regulate the grid companies' revenue caps and define a set of general rules for transmission and distribution tariffs. These regulations also define financial and technical reporting requirements to NVE. Section 7.4 describes the calculation of distribution and transmission tariffs. Section 7.5 describes the regulatory regime in more detail.

7.2.2.3. Governmental Regulations on Metering and Accounting. These governmental regulations define responsibilities and procedures related to metering and accounting, and include procedures and formats for information exchange between market participants. These regulations also define routines for switching suppliers. Procedures for handling small-customer access without remote hourly metering are described, including a model for load profiling in a system with several suppliers. Section 7.5 details these structures.

7.2.2.4. Guidelines for System Operations. These guidelines define the responsibility of Statnett concerning system operations and Statnett's and the grid users' rights and obligations regarding system services.

7.2.2.5. Guidelines for Energy Efficiency. These guidelines define the responsibility of the grid companies concerning energy efficiency. Among other things, the grid companies are required to inform the customers of possibilities for Demand Side Management (DSM) and energy efficiency, and provide historic meter values when necessary. Costs related to DSM and energy efficiency are covered through a separate charge added to the distribution tariffs.

7.3. THE WHOLESALE ELECTRICITY MARKET

The traditions for a wholesale electricity market go back to the early 1970s in Norway. The Norwegian Power Pool, Samkjøringen, was established to coordinate and optimize the output of the Norwegian hydroelectric system. This electricity market was mainly open to Norwegian generators, although a few large customers were also given access to the market (i.e., power-intensive industries). The pool price was based on marginal operating costs. The market was an interutility market for surplus power. The pricing principles were much the same as in the current power exchange.

When the electricity market was deregulated in 1991, the transmission grid was divested from Statkraft into a separate company—Statnett. Statnett Marked, the forerunner of Nord Pool, was established as a subsidiary of Statnett in 1993. As opposed to the former Samkjøringen, Statnett Marked was open to all market participants meeting the requirements of the power exchange, meaning that the demand side was included in the bidding process. Beyond voluntary trading in the organized market, market participants can enter bilateral contracts in the Over-the-Counter (OTC) market.

Other Nordic companies were given access to the spot market on special terms over the period 1993–1996. In January 1996, the geographic trading area of the power exchange was expanded to include Sweden in a joint Nordic power pool. Simultaneously, Statnett Marked changed its name to Nord Pool. The Swedish national power grid, Svenska Kraftnät, acquired 50% ownership of Nord Pool in April of the same year. Finland has operated a separate electricity exchange, EL-EX, since 1996. In June 1998, however, EL-EX joined Nord Pool, and Finland was established as a separate bid area. Svenska Kraftnät and Fingrid share the ownership of EL-EX. The Western part of Denmark—Jutland and Funen—was included as a separate bid area on July 1, 1999, and Zealand entered on October 1, 2000.

7.3.1. The Energy Markets

Nord Pool organizes two different markets for trade in electricity, Elspot and Eltermin, whereas Statnett operates the regulation market in Norway (Nord Pool, 1998). The total volume traded on Nord Pool's markets in 1999 was 216 TWh for Eltermin and 75 TWh for Elspot. The liquidity of Nord Pool's markets and products increased considerably during 1999. Additionally, Nord Pool's clearing service traded 684 TWh in 1999.

7.3.1.1. Elspot—The Spot Market. Power contracts for next-day physical delivery are traded in the spot market on an auction basis. A price per MWh is determined for each hour of the 24 hours in each one-day period. The participants' bids and offers are grouped together to form a supply curve (sale) and a demand curve (purchase). The system price (i.e., the unconstrained market-clearing price) is determined where the two schedules intersect.

7.3.1.2. Eltermin—The Futures and Forwards Market. Eltermin is a futures market for cash settlement of a specified volume of power at a negotiated price, date, and period. The market participants may trade in the futures market up to three years in advance. Futures are used for price hedging and risk management. Contracts traded on Eltermin are defined with a weekly resolution. Contracts due more than 5–8 weeks ahead are grouped in blocks of 4 weeks each. Contracts that are due for delivery more than one year ahead are combined in seasonal contracts: Winter 1 (weeks 1–16), Summer (weeks 17–40), and Winter 2 (weeks 41–52/53). The system price, i.e., the Elspot price, is used as a reference price for these contracts. Starting in October 1997, forward contracts were traded on Eltermin. The main difference between ***futures*** and ***forwards*** is the daily marking to market and settlement of futures. Forwards, on the other hand, are settled when the contract is due for delivery.

7.3.1.3. ELBAS—The Joint Swedish/Finnish Adjustment Market. Besides the two markets organized by Nord Pool, Finnish EL-EX organizes an intermediate market—the ELBAS market; it is a two-hour ahead market that opened on March 1, 1999. The ELBAS market offers the Swedish and Finnish participants the opportunity to adjust schedules closer to the hour of operation. The products traded at the ELBAS market are 7 individual hours at a minimum, and 31 individual hours at a maximum. The price per MWh is determined through continuous trade; the electronic trade system PowerClick is open to trade 18 hours a day, and bids are submitted and ranked based on price and, to some extent, time. A trade is closed when purchase and sales prices meet, and the trade is automatically transferred to clearing.

7.3.1.4. The Regulation Market (Imbalance Market). Statnett operates the regulation market to secure real-time balance between generation and load. Active bidders on the regulation market must regulate their delivery and usage within 15 minutes notice. So far, only generators have submitted bids to this market, after which Statnett sorts the bids in merit order for each hour. Generators are called upon to adjust the balance when necessary. Recently, the demand side has been involved in real-time balancing. Statnett has entered contracts with power intensive industries for real-time adjustment of load. However, this solution is not satisfactory, particularly for smaller consumers. Demand side bidding is under consideration for all customers. Market participants, other than the price setter, are price takers in the regulation market. So, they are charged the ex-post price for the imbalance between their scheduled and metered loads.

Svenska Kraftnät is responsible for real-time balancing between generation and load through the balance market in Sweden. Bids for regulation of generation and loads are arranged in merit order and Svenska Kraftnät uses the bids in the operational phase in the same way that Statnett does in the regulation market. The Finnish regulation market, which maintains the momentary balance between generation and load, is run by a subsidiary of Fingrid. The Finnish regulation market is different from the Norwegian and Swedish markets in that the participants can buy a right to deviate from a balanced schedule. The operator of the regulation market is defined as an open supplier covering the deviation. Eltra and Elkraft are responsible for the regulation markets of Denmark.

7.3.2. Zonal Pricing

Congestion in the Nordic power system is managed through a combination of zonal pricing for day-ahead congestion and countertrades (a buy-back model) for real-time congestion. Nord Pool is responsible for calculating zonal prices for congestion on major transmission lines. Norway is divided into several zones, whereas Sweden, Finland, and Denmark are separate zones. Statnett defines the Norwegian zones for 6 months. Market participants submit bids and offers in these zones. After calculating the system price (the unconstrained market-clearing price), the power balance in each price area is considered. When the system price is settled, the SO knows the total generation and load within each zone, and the net surplus or deficit in the different areas is found. If the net position in different areas indicates an overload on transmission lines between areas, the prices are adjusted to keep the transfer within capacity limits. The price in surplus areas is reduced, whereas the price in deficit areas is increased relative to the system price (the method is described in more detail in Section 7.6.2). This calculation of zonal prices is based on adjustment bids submitted by market participants in the day-ahead market. Nord Pool currently manages the capacity charge settlement, i.e., the fee market participants must pay as a response to zonal prices. Revenues from this settlement are transferred to Statnett, which reduces the size of the demand charge of the transmission tariff.

Real-time congestion is handled through countertrade or a buy-back model. The SO buys decremental or incremental regulation on each side of the congested path to adjust real-time transmission to available capacity (Grande and Wangensteen, 2000).

7.3.3. Ancillary Services

NVE's Guidelines on System Operations regulate ancillary services in Norway. Statnett is responsible for operating ancillary services in the Norwegian power system. Except for secondary control, which is traded in the regulation market, ancillary services are provided through contracts with users of the transmission grid rather than through a market. Generators connected to the transmission grid have an obligation to provide ancillary services, only limited by the technical capacity of their equipment. The generators will receive reimbursement only if the ancillary service is required beyond the prenegotiated limit. This compensation is based on installed capacity, available reserve (stand-by), and activated reserves.

The ancillary services are defined by NORDEL as follows:

- *Primary Control* is spinning reserve that is automatically begun; it is used for frequency control and contingency reserve and requires response within 30 seconds.
- *Secondary Control* is manually activated reserve used to regulate area control errors and time deviations; it requires accessibility within 15 minutes and is provided through the regulation market in Norway.
- *Reactive Power* is a reserve that is automatically activated when voltage deviations occur.
- *Generation Tripping* is predefined disconnection of production when a special operational disturbance occurs.
- *Load Shedding* is predefined disconnection of load when frequency decreases.

7.3.4. Bilateral Trading

Approximately 90% of all wholesale trade takes place in bilateral contracts, or within vertically integrated utilities. Both financial and physical bilateral contracts are traded OTC in Norway. Physical contracts imply physical delivery of the electricity traded. Financial contracts, on the other hand, imply cash settlement; no physical delivery takes place.

Trade in physical bilateral contracts between two defined price zones is treated the same way as spot power, i.e., "scheduled" with the Nord Pool. The reason for this is that Statnett needs a continuous overview of the total power flow to manage congestion. Unless the market participant has other ways of balancing obligations and rights (netting), bilateral contracts are scheduled in one of two ways:

- If the sale is within one other price zone only (either seller or buyer schedules the physical trade), the seller schedules the sale as a purchase in the receiving zone or the buyer schedules the purchase as a sale in the zone of generation.
- If the sale is between two different nonlocal areas (purchaser's receiving zone differs from the seller's zone of delivery): Both parties must take the contractual quantity into account when scheduling bids and offers.

In April 1997, Nord Pool started offering bilateral clearing in addition to the Eltermin clearing service. Clearing implies that Nord Pool is the legal counterparty with both buyer and seller of a bilateral contract. Clearing of contracts removes the counterparty risk in the hands of the parties entering a contract. Nord Pool takes the risk, but is secured through margin accounts and margin calls. Companies registered at the Eltermin (and broker participants) can use bilateral clearing. Bilateral clearing is offered for all Eltermin-type derivatives. When clearing bilateral forward contracts, three reference prices are currently allowed: (1) the unrestricted system price, (2) the Smestad price (for the Oslo area), and (3) the Stockholm price.

7.4. TRANSMISSION/DISTRIBUTION ACCESS, PRICING, AND INVESTMENT

Allowing third party access and implementing transmission-pricing principles to ease free trade are important in a competitive electricity market. The traditional Norwegian transmission tariff for firm power, with a distance element depending on who traded with whom, did not fulfill this requirement. The Point of Connection Tariff was therefore introduced as a general principle for transmission and distribution in May 1992.

7.4.1. Overall Principles: The Point of Connection Tariff

The basic principle of the Point of Connection Tariff is that each grid user (end-users, generators, or other grid owners) pays a transmission tariff depending on the point of connection. The transmission tariff in each point of connection is calculated relative to a defined, fictitious "marketplace" in the central grid. The seller pays for the electricity being transported into the marketplace, whereas the customer pays for transport out of the marketplace. Figure 7.3 shows the structure of the Norwegian Point of Connection Tariff.

The grid users face only one tariff, i.e., the tariff from the central (transmission), regional, or local grid levels. Costs related to higher-voltage (superjacent) grid lev-

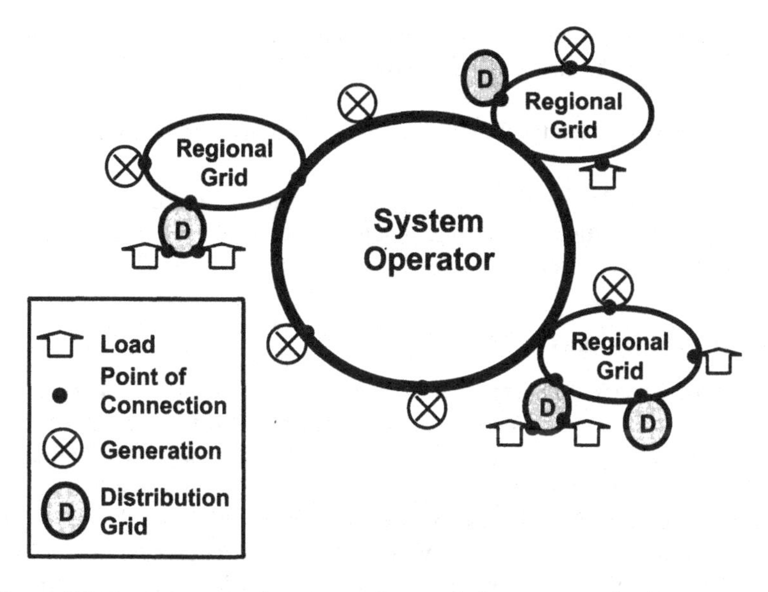

Figure 7.3. Interconnection between total transmission system and grid customers.

els are passed through to grid users connected to the grid level in question. Every network operator must prepare transmission tariffs, which are based on the following principles:

- The network operator must define points of connection where exchange (injection or withdrawal) of power with others (generators, end users, or other network operators) is done.
- Tariffs refer to these points of connection.
- Revenues from tariffs should be sufficient to cover costs related to individual networks and tariffs paid to superjacent grids, within the permitted revenue. When the network operator is efficient, transmission tariffs should provide a reasonable return on invested capital.
- Grid users connected to the network need only one agreement to gain access to the power system. The user signs a contract with the connecting network.
- Transmission tariffs must be determined independently of power purchase agreements.
- Tariffs should stimulate efficient use of the network.
- Tariffs are required to be public and nondiscriminatory.
- A network agreement, generally, is between the distribution network and the connected customer.

The governmental regulations detailing calculation of tariffs are based on the principles of the Point of Connection Tariff. This theoretical base applies both to distribution and transmission. The regulations require the entry/exit tariffs for withdrawal and input to be structured as follows:

1. *Volume-dependent* tariff elements vary according to grid user's withdrawal or injection.
2. *Other* tariff elements do not vary according to the grid user's metered withdrawal or injection. The charge is supposed to be neutral to consumption, and is a residual charge.

7.4.2. Transmission Tariffs

The volume-dependent charge of the transmission tariff, i.e., the *energy charge,* must be geographically and periodically differentiated. Statnett calculates the *marginal loss percentages* through representative load flow simulations for the Norwegian/Swedish system. Each tie point (with a total of 171 points) in the Norwegian transmission grid has an individual marginal loss percentage attached to it. The absolute value for injection versus withdrawal is the same at each tie point (however, with different signs). The denominations will vary for load versus generation, depending on the balance between the two at the specific tie point. The charges are calculated for periods of 8 to 10 weeks, a fortnight in advance at the latest. The charge is differentiated seasonally and for day and night.

Grid owners are responsible for grid losses in Norway. These losses are bought in the market, and therefore reflect the spot price. Marginal costs of electricity transmission must be reflected in the volume-dependent part of the tariff. This normally implies that grid losses, reflecting varying physical losses and spot prices, are covered through a volume-dependent charge.

Whenever there is congestion between two zones, the volume-dependent charge will additionally include the *capacity charge,* which reflects the zonal price differences. The generators in surplus zones (where generation is greater than demand) face a positive capacity charge. The charge is the difference between the zonal price (as calculated by Nord Pool) and the system price (the unrestricted Elspot price). The generators in deficit zones (where generation is less than demand) will, on the other hand, face a capacity charge with negative denomination from the difference between the zonal price and the system price.

The transmission tariff consists of two other charges: one being the *gross charge* and the other the *net charge.* These charges are differentiated for injection and withdrawal. The hours with maximum load in the areas North, Middle, and South are used for settlement.

The gross charge is settled based on the peak load in winter at each tie point. End-users face a gross charge based on total withdrawal from the grid and suppliers face a gross charge based on total injection to the grid. The gross charge for generators is based on installed capacity.

The net charge is settled based on the grid user's net exchange with the transmission grid. The end-user is charged for net withdrawal from, whereas the supplier is charged for net injections to, the grid.

The difference between the gross charge and the net charge is the volume used for settlement. The gross charge is based on total withdrawal and/or injection to/from the grid, and the net charge is based on either net withdrawal or net injection.

Statnett is responsible for grid expansion, new investment, and maintenance of the transmission grid. Statnett can finance future investment in the grid in two ways: (1) through the general tariff elements described above or (2) through a *construction contribution*. The construction contribution is a payment charged to grid users benefiting from the grid investment in question, and is a payment made only once. In other words, besides paying the construction contribution, contributors pay the general tariff, as do all other grid users. However, in meshed grid structures the transmission company's possibility of charging a construction contribution is restricted.

On January 1, 1997, compensation for *Energy Not Supplied* (ENS) was introduced for grid users connected to the transmission grid. The concern was (1) that the tie points of a socially efficient grid can have different reliability and (2) that the system under some situations can be run with lower operational security than stated by the *n – 1 check criterion.* (The *n – 1 check criterion,* traditionally used when planning grid expansions, implies that the grid is built so that supply is secure even if one line or connection fails.) ENS is therefore expected to lead to investment decisions being more socially efficient than in previous years. Some important rules related to Statnett's 1997 use of ENS were as follows (Voldhaug, Granli, and Bygdås, 1998):

- ENS is only granted when it is the central grid that caused the disturbance.
- Outages lasting more than 3 minutes are compensated at U.S.$2330/MWh.
- The grid user is responsible for notifying Statnett of ENS at its tie in order to be compensated.
- Statnett's total payment related to ENS is limited to 2% of the company's revenue cap.
- No ENS is paid in case of *force majeure,* such as extreme weather conditions.
- The individual grid user's total annual ENS is limited to 25% of the total payment from the grid user to the grid owner.

Statnett introduced ENS on a trial basis, and stopped the project after one year because no adjustment in the annual revenue cap was granted for these extra costs. However, a system of adjusting the revenue caps for ENS was started for all grid levels above 1 kV in 2001.

7.4.3. Distribution Tariffs

In calculating distribution tariffs, the distribution network operators must differentiate between customers metered on maximum hourly load, and customers that are not (NVE, 1999). This will usually imply differentiation between residential/other small customers and commercial/industrial customers.

Customers *not metered* on maximum hourly load [MW] are charged an *energy charge* [NOK/kWh] and a *fixed charge* [NOK/Year] representing the volume-dependent and other tariff elements, respectively. The energy charge is, at a minimum, required to reflect marginal losses in the distribution network and in higher voltage grids. The fixed charge, on the other hand, is required to cover customer-specific costs at a minimum. Examples of customer-specific costs are costs related to metering, settlement, and invoicing. Additional costs not covered by the minimum requirement energy charge and fixed charge are to be split between the two charges, as the local network grid owner finds appropriate.

Customers *metered* on maximum hourly load [MW] are charged a *demand charge* [NOK/MW] *plus* the energy charge *and* the fixed charge. The requirements for the energy charge and the fixed charge are the same as for customers not metered on a maximum hourly load. The additional costs not covered are, however, covered through a demand charge. This demand charge is normally based on the customer's maximum load in one or more months of the year.

7.5. DISTRIBUTION NETWORK REGULATION AND RETAIL COMPETITION

7.5.1. Rate of Return Regulation, 1991–1997

The first regulatory regime being used in the deregulated Norwegian electricity market was Rate of Return (ROR) regulation. In the period 1993 through 1996, NVE each year determined a cap on the rate of return from the total capital employed. The allowed ROR was based on the general interest rate in Norway, adding a risk premium of one percentage point. The maximum rates of return attainable for 1993, 1994, 1995, and 1996 were 11%, 7%, 7.5%, and 7.5%, respectively.

If the actual ROR was larger than the allowed rate, excess revenue was transferred back to customers over a period of three subsequent years through lower distribution rates. Likewise, if the actual ROR fell short, distribution rates could be increased over a period of three subsequent years. But ROR regulation led to excessive investment. Furthermore, the grid companies were not exposed to financial risk from temporary fluctuations in revenue. During the period 1993–1995, the grid companies, as a whole, transferred about NOK 1.4 billion (approximately US$165M) back to the consumers from excessive profits (Grasto, 1997).

7.5.2. Incentive-Based Regulation Starting January 1, 1997

The incentive-based regulatory model combines revenue cap regulation with benchmarking and earnings sharing mechanisms (see Chapter 4, Section 4.3.2). The regulatory period lasts 5 years; the first period covered 1997–2001, and the second period started in January 2002. Initial revenue caps in 1997 were based on the grid companies' accounts from 1994 and 1995, whereas the 1996–1999 costs form the basis for the initial revenue caps of 2002. Revenue caps are adjusted annually for

- A general and an individual productivity improvement factor
- An inflation factor
- A growth factor for grid expansions

The following general formula for revenue cap regulation is used:

$$IT_{e,n+1} = IT_{e,n} \cdot \left(\frac{KPI_{n+1}}{KPI_n}\right) \cdot (1 - EFK_{n+1}) \cdot (1 + SF \cdot \Delta GF) \quad (7.1)$$

where

$IT_{e,n}$ is the revenue cap of year *n*, excluding grid losses
KPI is the consumer price index
EFK is the efficiency improvement factor
SF is the scale factor for new investment
ΔGF is the growth factor for new investments

Losses are added to the formula through multiplying the physical losses by the spot market price. Therefore, losses are not adjusted by the inflation factor. However, losses are adjusted for the productivity improvement factor and growth factor.

As in Equation (7.1), the revenue caps are revised annually for an *efficiency improvement factor*. This calculation of the efficiency factor is different for distribution companies, regional grid companies, and the transmission company, and has been used at different times for the various grid levels.

During the first year that the regulatory model was in effect, the efficiency factor was equal for all distribution companies, and was set at 2%. In a subsequent years, starting in 1998, the efficiency factor has been unique for each distribution company, based on calculation of the cost efficiency of the different grid owners. Regional grid companies and the transmission owner had a general efficiency factor of 1.5% in 1997 and 1998. Starting on January 1, 1999, individual efficiency factors were also used for the 50 to 60 regional grid owners and the transmission owner. Figure 7.4 summarizes the measured cost efficiency for the 198 distribution companies based on fiscal years 1994–1995.

Figure 7.4 illustrates how many of the Norwegian distribution companies belonged to the different efficiency categories in 1997. Approximately 10 distribution companies were measured as less than 70% efficient, implying that these companies have efficiency potentials of more than 30% compared with the most efficient distribution companies. Similarly, as many as 50 distribution companies have been measured as near 100% efficient, meaning that these companies are the most efficient. The 1997 Norwegian model for *Data Envelopment Analysis* (DEA) applied to the distribution companies had five output variables and four input variables, as shown in Table 7.9.

New individual efficiency requirements were calculated for the second regulatory period starting in 2002. In addition to using data from the 1996–1999 period, the

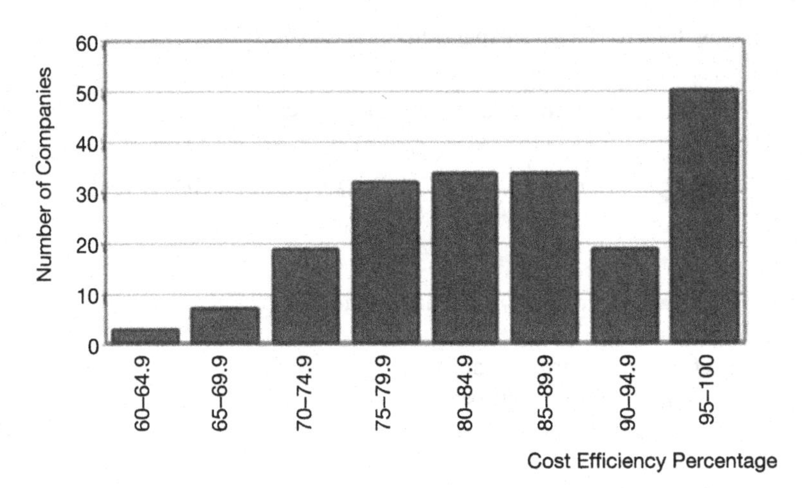

Figure 7.4. Distribution of measured cost efficiency for 198 Norwegian distribution companies. (Source: Norwegian Water Resources and Energy: www.nve.no.)

definition of input and output variables changed somewhat. A new input variable—actual costs of ENS—was introduced. The category "sea cables" was removed as a separate output variable and included in high-voltage lines and low-voltage lines, respectively. Additionally, expected ENS (NOK) was introduced as a new output variable.

In the DEA, factors describing outputs from inputs are calculated; for instance, capital investment per customer. With the variables in Table 7.9, 20 such factors are calculated for each distribution company. This is *technical efficiency.* To calculate *cost efficiency,* the input variables are priced and measured relative to the output variables. The factor prices used for "man years" and network losses are, respectively, the average salaries of each company per "man year" and average spot prices. The companies with least cost compared to the output variables are mea-

Table 7.9. Variables Used in the Initial DEA for Distribution Companies in Norway

Output Variables	Input Variables
Number of customers	Number of man-years
Energy delivered (MWh)	Network losses (MWh)
Length of higher-voltage lines (km)	Capital costs (NOK 1000), based on book or replacement value
Length of sea cables (km)	Services and goods (NOK 1000)
Length of lower-voltage lines (km)	

Source: Norwegian Water Resources and Energy: www.nve.no.

sured as 100% efficient and are used as references to measure the efficiency of the other companies (Kittelsen, 1993).

The DEA analysis is used to calculate the efficiency factors of the regional grid companies as well. However, different input and output variables are used: (1) input variables include "man years," grid losses, capital costs, "goods and services," and the actual costs of ENS and (2) output variables include maximum hourly load (MW), length of power lines, exchange to other grids, central grid facilities, as well as expected ENS (NOK).

The efficiency of Statnett—the Norwegian transmission owner and system operator—is calculated separately (ECON, 1999). Statnett originally had two revenue caps: One revenue cap related to its role as system operator and another related to managing the transmission grid. The revenue cap of Statnett as the transmission owner is covered in an efficiency study. The efficiency of Statnett's transmission grid management is calculated in two steps:

1. The efficiencies of grid construction and operations and maintenance (O&M) are calculated separately.
2. The total efficiency of Statnett as a transmission company is calculated as a weighted average of the efficiency of grid construction and the efficiency of O&M.

The Swedish SO and transmission owner—Svenska Kraftnät (SK)—has been used as a benchmark to calculate the efficiency of Statnett in the first regulatory period. The efficiency of Statnett has been calculated by comparing the relations between cost (*C*) and cost drivers (*CD*):

$$Efficiency_{Statnett} = \frac{\left(\dfrac{C_{SK}}{CD_{SK}}\right)}{\left(\dfrac{C_{Statnett}}{CD_{Statnett}}\right)} \tag{7.2}$$

Statnett was measured as 74% efficient compared with Svenska Kraftnät. Svenska Kraftnät might not be 100% efficient though, implying that the potential for efficiency improvements of Statnett might be even larger than 26%.

The growth factor for grid expansions in the Norwegian model was originally based on the parameter load growth multiplied by a scale factor of 0.5. This implied that if a grid company had a revenue cap of US$1M the previous year, and a load increase of 2% was expected, the revenue cap was increased by US$10,000 to cover necessary grid expansions. However, load growth as a measure of growth has several disadvantages (Jordanger and Grønli, 2000). Therefore, the regulator has made some changes to the mechanisms to adjust for grid expansions for the second regulatory period. The foreseen growth factor is a combination of average national load growth (with a substantially smaller scale factor) and the relative growth of new buildings in the grid company's supply area. The final scale factors to be used will be determined during 2002, and adjustments will be made ex-post as opposed to ex-ante in the first period.

The revenue caps are combined with an earnings-sharing mechanism, i.e., there is a maximum and a minimum limit for allowed profit. The maximum limit for profit is to protect the customers from unreasonably high prices, and was set at 15% of invested capital in the first regulatory period, and increased to 20% in the second period. The grid owners are guaranteed a minimum rate of return of 2% of invested capital to ensure a minimum standard and quality.

7.5.3. Retail Competition—Important Developments

In Norway, retail competition was introduced in 1991. There have, however, been some important changes over the years that heavily influenced the frequency of customers changing suppliers. The number of residential customers with other than an incumbent electricity provider at different dates, including January 1, 2001, can be seen in Figure 7.5. (The number of customers with suppliers other than the local one is registered quarterly by NVE.)

Not many residential customers shopped around in the period 1992 through 1996. Before 1997, only 0.05% of the residential customers had switched electricity supplier. There are several explanations for the development shown in Figure 7.5:

- Suppliers had to pay a fee of U.S.$500–670 for serving customers in areas other than their local area until 1995, when this fee was removed.

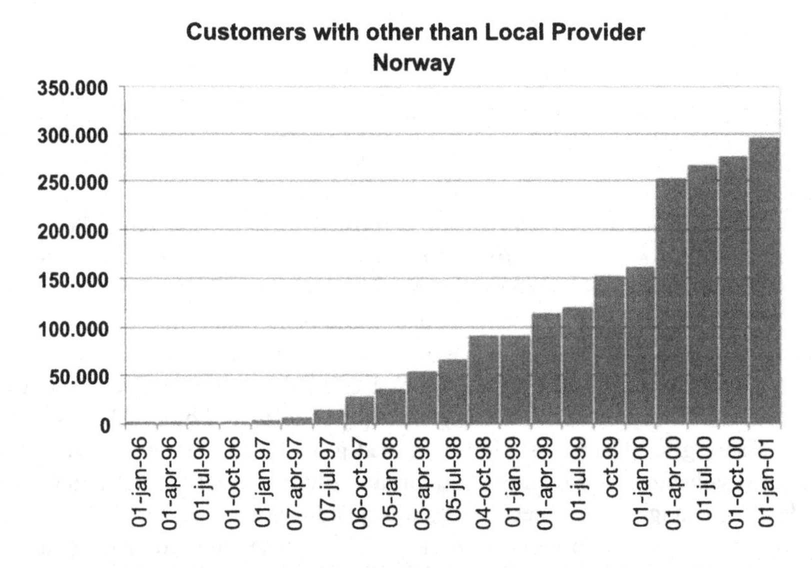

Figure 7.5. Residential customers with other than incumbent electricity provider, January 1, 1996 through January 1, 2001. (Source: Norwegian Water Resources and Energy.)

- Load profiling was introduced to help the change of supplier for small customers in 1995. Remote metering systems (hourly) are not required for customers with annual consumption below 400 MWh (this was changed from 500 MWh in January 1999).
- Customers had to pay a switching fee to grid owners of approximately U.S.$30 until 1997.
- NVE required suppliers/grid owners to facilitate more frequent switching in 1998. Customers may now switch supplier each week on 3 weeks notice. Quarterly invoices based on metered values are required.

After 11 years of competition in the electricity industry in Norway, the number of residential customers affiliated with suppliers other than the local supplier is about 17.7%. If the average residential customer consumes 20 MWh annually, customers with a different supplier than the local one represent a load of approximately 8087 GWh a year. In addition, about 65,500 commercial customers had a different electricity provider than the local one in September 2002, representing 22% of all commercial customers.

Market access without hourly metering systems is possible for small customers through a method of estimated load profiles for specific areas, rather than customer-specific load profiles in invoicing and settlement (Livik and Fretheim, 1997). Figure 7.6 illustrates this method.

Network operators are responsible for submitting meter reading data to all electricity providers serving customers in the local area. Since small customers are not metered hourly, a method for distributing the metered load among different suppliers' customers must be applied. The first step is to find the total load to be distributed among the electricity suppliers. This is done by subtracting grid losses and hourly metered consumption from the area's total load profile. The resulting load profile is distributed among the electricity suppliers having customers that are to be settled by profiling. The second step is to use the last period's meter data (year/quarter/month) to calculate each electricity provider's share of the load profile (as a percentage of the total load profile of the local grid company). In other words,

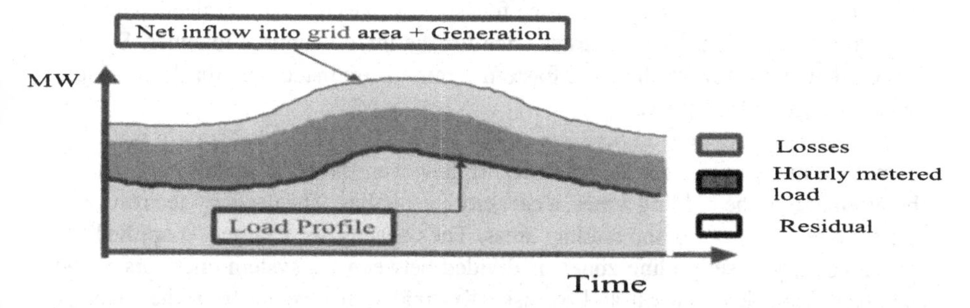

Figure 7.6. Calculating the load profile for settling customers without hourly remote meter systems.

customers settled by load profiling are distributed among electricity providers. However, because there are mutual differences in the relative load patterns (and varying consumption patterns) for each customer, relying on the load profile for settlement would be incorrect over longer periods. The network operator, therefore, must read the meters regularly to make final settlements among the electricity providers of the concession area.

The method of load profiling was an important element in providing retail access for small customers. However, the importance of the method will be reduced by increasing the frequency of meter reading and by expanding two-way communication installations and hourly meter reading systems.

7.6. ASPECTS OF THE REGULATORY PROCESS IN NORWAY

As opposed to regulatory processes in other countries, deregulation did not require electricity rates to be reduced. Securing market-based rates was assumed to lead to rate reductions. Market participants were also not allowed recovery of stranded costs when the market was opened to competition.

7.6.1. The Inter Nordic Exchange

As of January 2000, Nord Pool had 264 registered market participants from Norway, Sweden, Finland, Denmark, and England and Wales. Some of these participants do not trade on their own account with the Exchange, but are registered as so-called clearing customers or broker participants.

The joint Nordic exchange has defined bidding areas to which market participants can apply their bids. Sweden, Finland, and Eastern and Western Denmark are individual bid areas, whereas Norway has several predetermined bid areas. Swedish market participants refer their bids to Zone A–Sweden. Finnish market participants refer their bids to Zone B–Finland. Danish market participants refer their bids to Zone C or D. The internal zones in Norway are defined by the SO for 6 months, and are called E, F, etc. Norwegian participants refer their bids to the respective Norwegian zone. Figure 7.7 illustrates the bidding areas applying to Nord Pool.

Nord Pool calculates a system price for Norway, Sweden, and Finland. The system price is the unrestricted market price for the three countries together. System price calculations for Western and Eastern Denmark are made only for the available capacity used in Elspot trading.

If there is no congestion between any of the bidding areas, the system price becomes the market price for the countries. If, however, there are capacity constraints between any of the bidding zones, a capacity fee applies. The capacity fee results in different zonal prices in the bidding areas. The capacity fee charged, if applied between country-wise bidding zones, is divided between the system operators of the involved countries. Statnett and Svenska Kraftnät split revenue from the capacity charge with restrictions between Norway and Sweden. Svenska Kraftnät and Fingrid split revenue with restrictions between Sweden and Finland.

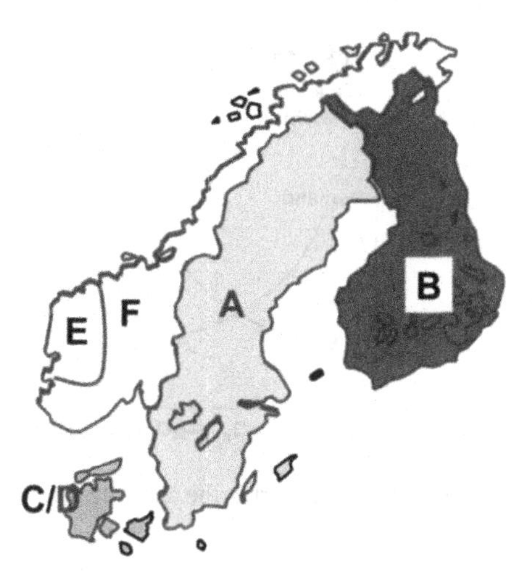

Figure 7.7. Nord Pool's bidding areas.

All market participants can bid in their local currency. Nord Pool offers a free exchange service that makes it possible for Danish, Finnish, and Swedish participants to receive settlement in DKK, NOK, SEK, and EURO. Settlement is made in the currency in which the bid was submitted. The same trade rules and fees apply to Danish, Finnish, Norwegian, and Swedish participants. Wheeling in/out transmission tariffs do not apply to electricity exchange cross the borders between Finland, Norway, and Sweden. There are, however, border tariffs to Western and Eastern Denmark. When bidding for Elspot or Eltermin, Norwegian participants submit bids to Nord Pool's office in Oslo. Danish, Finnish, and Swedish participants submit the bids to Nord Pool's office in Stockholm. For bidding into the ELBAS market, Finnish and Swedish participants submit bids to the Nord Pool/EL-EX office in Finland.

Border tariffs between Norway and Sweden were removed in 1996 when the joint Norwegian/Swedish pool was established. The border tariffs between Finland and Sweden were removed on March 1, 1999, simultaneously with the launching of the ELBAS market. Consequently, no wheeling (through or in) tariffs remain within the deregulated Scandinavian electricity markets, except for trade with Denmark. Furthermore, there are differences among the three countries regarding how the principles of the Point of Connection Tariff are implemented.

7.6.2. Congestion Management in the Scandinavian Area

Congestion management for the combined deregulated Scandinavian markets is solved through zonal prices calculated by Nord Pool. The three countries are separate zones, and congestion between the country zones is solved through capacity charges. The *capacity charge* (CC) yields different prices in the three countries

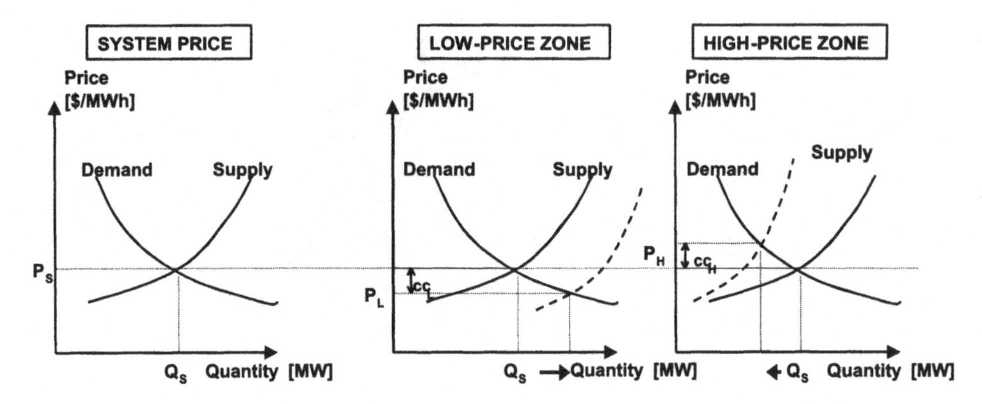

Figure 7.8. Calculating the capacity charge.

when the interfaces between the zones are congested. Figure 7.8 illustrates how the capacity charge is determined.

At the market solution of the unconstrained market-clearing price (the system price) and the quantity Q_s, there will be congestion (Figure 7.8). To relieve the congestion at the interface between the two areas, capacity charges are used. The net demand in the zone to which there is congestion must be reduced. A capacity charge of CC_H for suppliers wanting to supply this area is introduced. Alternatively, a negative capacity charge of CC_H to the companies demanding electricity in the area pushes the zonal price upwards until the area quantity is reduced to a level where transfer to the area is uncongested. Likewise, the net supply of the zone from which there is congestion must be increased to relieve the constraint. A capacity charge of CC_L on demand in the zone is introduced. Alternatively, a negative capacity charge of CC_L on supply in the zone pushes the zonal price down until the area quantity is increased to a level at which transfer from the area is uncongested.

Congestion within the country zones, however, is solved differently in each individual country. Statnett uses capacity charges for congestion inside the country borders as well. Svenska Kraftnät and Fingrid both use the "buy-back principle" for congestion management within Sweden and Finland, respectively. This implies that Svenska Kraftnät and Fingrid choose the least expensive units from a country-specific merit order list of adjustment bids until the internal congestion is relieved. This merit order list is the same as the merit order list of the regulation market.

CHAPTER 8

THE SPANISH POWER SECTOR

8.1. GENERAL DESCRIPTION OF THE SPANISH POWER SYSTEM

In 1998, the Electricity Law 54/1997 introduced a new configuration for the Spanish electricity system. Before this law, in December 1996, the Ministry of Industry and Energy signed the *Electricity Protocol* with the electric utilities outlining the general structure of these changes. Broadly speaking, the *Electricity Protocol* called for the transformation of the Spanish system from a central purchasing agent model to a model of wholesale and retail competition.

8.1.1. Structure of the Industry

Before restructuring, most of the Spanish electricity companies were vertically integrated. Electricity Law 54/1997 required accounting and legal separation of regulated activities (i.e., transmission and distribution) from nonregulated activities (i.e., generation and retail). Accounting separation was started immediately, whereas the legal separation was required before the end of the year 2000. Retail competition was established through the creation of new retail companies that could sell energy to *qualified* customers, i.e., nonregulated customers that could choose their supplier. The Electricity Law established a 10-year transition schedule to gradually introduce customer choice.

8.1.2. Generation

Before restructuring, there were six major electricity companies: Iberdrola, Endesa Holding (with Enher, Hecsa, ERZ, Viesgo, Gesa, and Unelco), Unión Fenosa, Hidrocantábrico, Sevillana, and Fecsa. Most of them were vertically integrated with generation and distribution assets. These companies, except Endesa Holding, were private. In 1997, Endesa Holding purchased Sevillana and Fecsa, and in 1998 Endesa Holding was also privatized. Since the competitive wholesale Spanish market was opened in 1998, the four generation companies (Endesa Holding, Iberdrola, Unión Fenosa, and Hidrocantábrico) have been selling energy in the market on a daily basis.

Table 8.1 presents the installed generation capacity and the annual energy production by generation technology. Table 8.2 shows the energy generated and sold by Spanish generation and distribution companies.

Table 8.1. Installed Capacity and Annual Energy Production by Technology in 1998

	Capacity	Energy
Coal	26%	35%
Cogeneration and renewables	NA	8%
Hydro	38%	20%
Nuclear	17%	34%
Oil/gas	19%	3%
Total	43.5 GW	173 TWh

Source: Red Eléctrica de España: www.ree.es.

8.1.3. Transmission

Since 1984, under the previous regulatory framework, the transmission system was owned and operated by *Red Eléctrica de España* (REE). The Electricity Law consolidated REE as the System Operator (SO) and transmission owner, and separated the Market Operator (MO) functions into a new company, *Compañía Operadora del Mercado Español de Electricidad* (OMEL). REE is the principal Spanish power transporter and network operator and controls technical dispatch and international exchanges. REE was initially state-owned, but the government has divested part of its shareholding. The law established that any shareholder should not own more than 10% of the total share capital, and the sum of wholesale market agent shares should be lower than 40%. Table 8.3 shows the transmission facilities owned by REE and the rest of utilities. Tables 8.4 details the main interconnection lines with neighboring countries. Table 8.5 gives net energy exchanges between Spain and its neighbors.

Table 8.2. Generated and Sold Energy by Companies in 1996

Utility	Generated Energy	Sold Energy
Endesa	30.1%	0.0%
Enher and Hecsa	3.5%	10.2%
ERZ and Viesgo	1.8%	5.6%
Sevillana	6.3%	14.7%
Fecsa	7.9%	11.0%
Iberdrola	32.6%	38.6%
Union Fenosa	12.1%	15.5%
Hidrocantabrico	5.0%	4.4%
Other	0.7%	0.0%
Total	142 TWh	142 TWh

Source: Red Eléctrica de España: www.ree.es.

Note: Cogeneration and renewables have not been included. Also, GESA and UNELCO utilities, which operate respectively in the Balearic and Canary Islands, have not been included.

Table 8.3. Transmission Facilities in 1997

	REE	Rest of Utilities	Total
400 kV Lines	98%	2%	14,244 km
220 kV Lines	27%	73%	15,701 km
400/HV Transformers	40%	60%	42,687 MVA

Source: Red Eléctrica de España: www.ree.es.

8.1.4. Distribution

Table 8.6 presents distribution network installations ranging from high voltage (HV), to medium voltage (MV), to low voltage (LV). Table 8.7 shows distribution companies that cover the Spanish territory.

8.1.5. Consumption

Regarding energy consumption, 53.7% of all electrical energy is supplied through the high-voltage and medium-voltage networks, 23.7% of total energy is for residential low-voltage consumption, and the rest (22.6%) is for other low-voltage uses. Table 8.8 shows the energy consumption and the number of customers connected at each distribution voltage level.

Electricity Law 54/1997 introduced retail competition progressively by allowing

Table 8.4. Interconnection Lines and Commercial Interconnection Capacity

	France	Portugal	Morocco
Number of lines	6	7	1
Voltages	2 × 400 kV	2 × 400 kV	400 kV
	2 × 220 kV	3 × 220 kV	
	2 × 132 kV	1 × 132 kV	
		1 × 66 kV	
Maximum capacity	3270 MVA	4255 MVA	700 MVA
Commercial capacity (imports)	1100 MW	650 MW	300 MW
Commercial capacity (exports)	1000 MW	750 MW	350 MW

Source: Red Eléctrica de España: www.ree.es.

Table 8.5. Net Energy Exchanges in GWh: (+) Imports; (–) Exports

Year	France	Portugal	Andorra	Morocco	Total
1996	2,291	–1,106	–125	—	1,060
1997	27	–2,897	–105	–131	–3,106
1998	4,519	–277	–152	–706	3,384

Source: Red Eléctrica de España: www.ree.es.

Table 8.6. Distribution Installations in 1995

Distribution Installations	Total
HV Lines (36 kV < V < 220 kV)	46,367 km
Number of HV/MV substations (capacity)	1,434 (59,147 MVA)
Number of MV/MV substations (capacity)	731 (4,454 MVA)
MV Lines (0.38 kV < V < 36 kV)	209,660 km (83% overhead)
Number of MV/LV substations (capacity)	257,687 (74,421 MVA)
LV lines	281,713 km

Source: CNSE (1995).

Table 8.7. Percentages of National Territory Covered by Distribution Utilities

	Percentage of National Territory
Endesa Holding (including Sevillana and Fecsa)	41%
Iberdrola	39%
Union Fenosa	16%
Hidrocantabrico	4%

Source: Red Eléctrica de España: www.ree.es.

some customers (qualified customers) to access the market directly or to buy energy from any retail company. Later regulations have increased the speed of market liberalization. Figure 8.1 presents the evolution of the number of customers that became nonregulated and the associated consumed energy during 1999. At the end of 1999, the number of nonregulated customers was 8000 of a total of 10,083 qualified customers.

8.1.6. Concentration Levels and Economic Indices

When the Endesa Holding bought Sevillana and Fecsa in 1997, market concentration in generation and distribution increased significantly. When the wholesale market started in 1998, market concentration reached the levels shown in Table 8.9. Two companies controlled almost 80% of the market with a high level of vertical integration between generation and distribution–supply activities. Table 8.10 shows economic indices and the evolution of tariffs for the Spanish electric sector.

Table 8.8. Energy Consumption and Number of Customers in 1997

Voltage Level	Energy Consumption	Number of Customers
High (36 kV < V < 380 kV)	33,862 GWh	1,082
Medium (1 kV < Voltage < 36 kV)	44,389 GWh	59,175
Low (<1 kV)	67,853 GWh	19,497,237

Source: UNESA (1998).

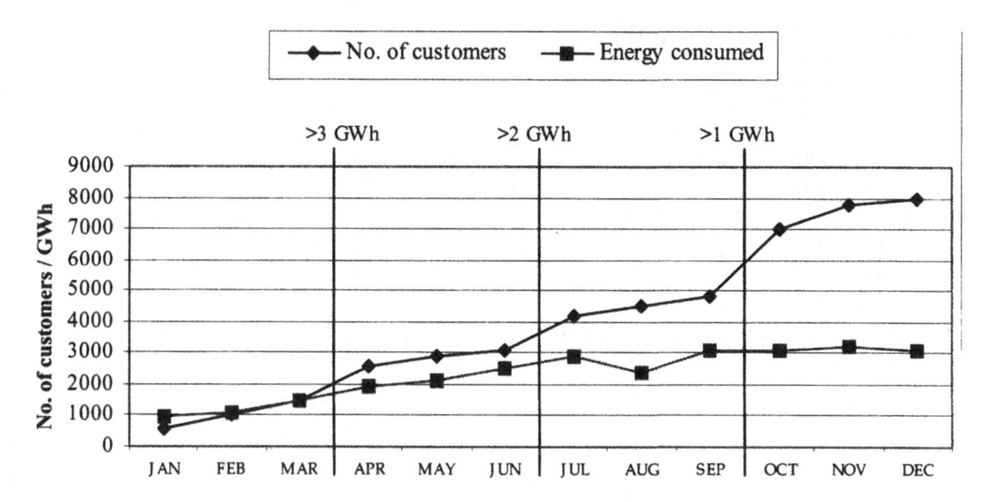

Figure 8.1. Evolution of the number of and energy consumed by nonregulated customers in 1999. *Source:* Comisión Nacional de Energía, CNE (2000).

Table 8.9. Concentration Levels in Competitive Generation and Distribution, 1998

	Generation (Average Production)	Distribution
Endesa Holding	50%	42%
Iberdrola	26%	37%
Union Fenosa	12%	15%
Hidrocantabrico	5%	4%
Other (Cogeneration and renewable)	7%	2%

Source: Red Eléctrica de España: www.ree.es.

Table 8.10. Annual Changes in Electric Sector Indices, 1988 to 1997

Year	Load Demand	Annual Incomes	Tariff	RPI(*)
1988	109 TWh	US$8.4B	5.50%	5.80%
1989	6.30%	10.05%	4.10%	6.90%
1990	3.63%	10.12%	5.50%	6.50%
1991	4.58%	11.80%	6.80%	5.50%
1992	1.24%	4.20%	3.20%	5.30%
1993	–0.36%	2.49%	2.90%	4.90%
1994	4.05%	5.17%	2.06%	4.30%
1995	3.75%	3.77%	1.48%	4.30%
1996	2.88%	2.86%	0.00%	3.20%
1997	(146 TWh) 3.56%	(13.6 B) 0.00%	–3.00%	2.20%

Sources: Red Eléctrica de España and UNESA (1998).
Note: Original data in Pta. Converted to U.S.$1 = 150 Pta.
*RPI = Retail price index annual variations.

Table 8.11. Evolution of the Different Costs

Cost Concepts	1988	1997
Fixed investment in generation	41.1%	27.3%
Fixed and variable O&M in generation	11.4%	11.3%
Fuel in generation	20.3%	22.7%
Transmission network owned by REE	2.2%	2.4%
Distribution	21.3%	24.7%
Utilities management	3.9%	3.5%
Specific taxes: nuclear industry, systems in Spanish islands, national coal, R&D	6.2%	11.0%
Other incomes	–6.2%	–2.9%
Total (*)	US$8.4B	US$13.6B

Source: UNESA (1998).
Note: Original data in Pta. Converted to U.S.$1 = 150 Pta.

The tariffs established under the earlier regulatory framework, known as *Marco Legal Estable,* are based on annual costs, shown in Table 8.11. After the establishment of the new electricity market, the actual average reduction in 1998 was about 3.63%. In September 1998, the Ministry agreed to minimum average reductions for the following three years: 2.5% in 1999, and 1% in 2000 and 2001. A different percentage reduction was to apply to different tariff categories. However, those eligible customers who chose to remain in the tariff did not see any tariff reduction until 2002.

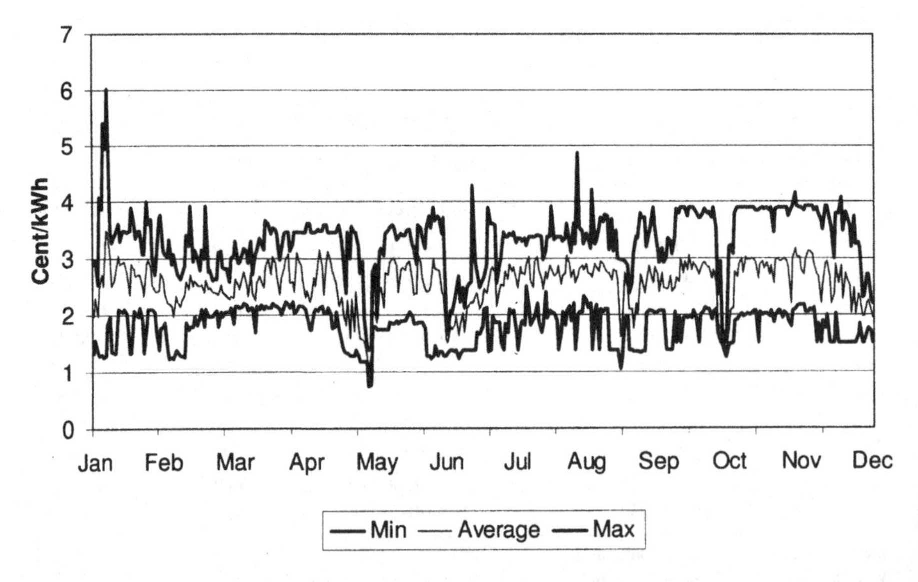

Figure 8.2. Evolution of final daily energy prices in the wholesale market in 1998. *Source:* OMEL (1998b); 1 $U.S. = 150 Pta.

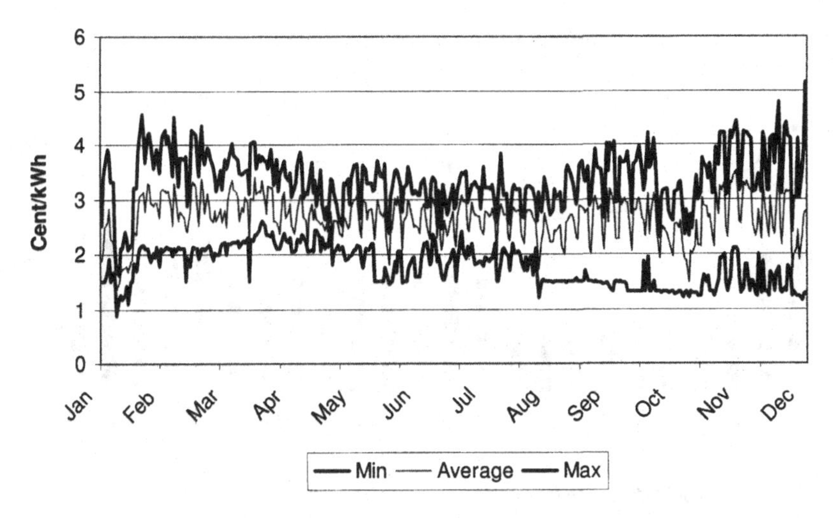

Figure 8.3. Evolution of final daily energy prices in the wholesale market in 1999. *Source:* OMEL (1999); 1 $USA = 150 Pta.

The evolution of final energy prices in the daily energy market in 1998 and 1999 is shown in the Figures 8.2 and 8.3. Approximately 75% of the final price corresponds to the energy in the day-ahead market, 3% to constraint management and ancillary services markets, and 22% to capacity payments.

Hourly prices in the day-ahead energy market are correlated with demand. Figures 8.4 and 8.5 show the evolution of prices for a weekday in July and a Sunday in January (randomly chosen).

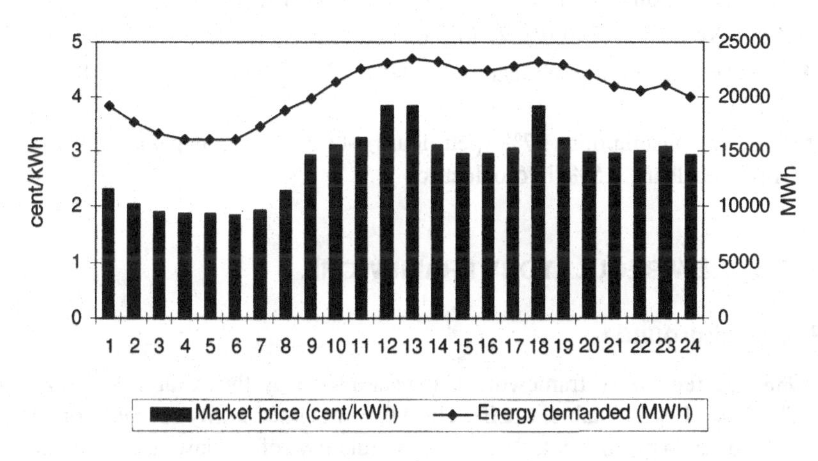

Figure 8.4. Evolution of hourly prices in the day-ahead energy market on a Wednesday. *Source:* Compañía Operadora del Mercado, www.omel.es.

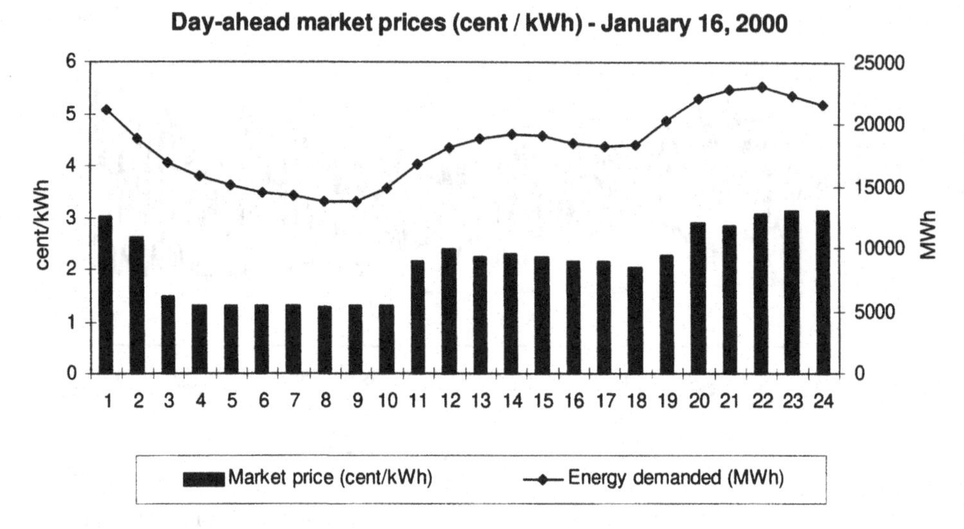

Figure 8.5. Evolution of hourly prices in the day-ahead energy market on a Sunday. *Source:* Compañía Operadora del Mercado, www.omel.es.

8.1.7. General Economic and Energy Indices for Spain (*Source:* http://www.eia.doe.gov/ emeu/cabs/)

- Population (2001E): 40 million
- Size: 0.5 million square kilometers
- Gross domestic product (GDP, nominal, 2001E): U.S.$579B
- Real GDP growth rate (2001E): 2.6%
- Inflation rate (2001E): 3.7%
- Oil production (2001E): 21,000 barrels per day (bbl/day)
- Coal Production (1999E): 27 million short tons
- Energy consumption per capita (1999E): 132.6 MBtu (versus 355.8 MBtu in US)
- Energy consumption: 57% (petroleum), 11.2% (natural gas), 14.3% (coal), 10% (nuclear), 7.5% (hydroelectric)

8.2. THE NEW REGULATORY FRAMEWORK

8.2.1. Background

In 1988, the regulatory framework was established by the "Stable Legal Framework" (Marco Legal Estable, MLE). The MLE regulated utility revenues on the basis of standard costs. In 1994, there was a regulatory reform law (Ley Orgánica del Sistema Eléctrico Nacional—LOSEN). Its objective was to introduce a greater degree of competition into the electricity sector, superseding the MLE. However, many of the arrangements foreseen in the LOSEN were not finally instituted. In January 1998, the Electricity Law 54/1997 introduced a new configuration for the Spanish electricity system. Before this law, the Ministry of Industry and Energy signed the Electricity Protocol with the electric utilities in December 1996 outlining the general structure of these changes. Table 8.12 summarizes the main milestones in the recent evolution of the Spanish electrical sector regulation.

8.2.1.1. Creation of REE. In 1984, the Spanish government created a new company under state control, Red Eléctrica de España (REE), to own and operate the transmission system. It was the first attempt to improve the overall efficiency in the sector by central coordination of all available resources and through central planning of new investments.

8.2.1.2. The Marco Legal Estable. The MLE was the regulatory framework from December 1987 (Royal Decree 1538) to December 1997. It determined the remuneration of electric companies using a cost-based approach. The main objectives pursued by the MLE were

- To promote the efficiency of the electric sector by means of a standard cost mechanism
- To stabilize the annual variation in electricity tariffs
- To warrant the investment recovery of fixed assets during an installation's operational life

Table 8.12. Evolution of the Electrical Sector Regulation in Spain

1984	National Unified Operation Law	Creation of REE National Grid Company Central Dispatch
Dec.1987	New Legal Framework Marco Legal Estable (MLE)	 Financial Stabilization Utility revenues based on standard costs National tariff system
Dec. 1994	Electricity Act (LOSEN)	First liberalization attempt Creation of the Regulatory Commission Open access to new entrants
Dec. 1996	Electricity Protocol (New Government-Utilities)	Basic principles for an overall competitive model Competition in generation Stranded costs recognition Reduction path for the regulated tariff Development of a new Electricity Law
Nov. 1997	New Electricity Law	
Jan. 1998	Implementation of the new electricity market model	

- To decrease uncertainty in the electric sector
- To adequately redistribute the incomes of the electric sector among private and public utilities
- To minimize the cost of service for final customers

The MLE system remunerated generation and distribution activities of each utility based on the cost of service and using standard cost coefficients. All customers under this national integrated remuneration system were price regulated. There was no competition in generation and customers were not allowed to choose among different suppliers. Each year, the national tariff was set, determining the total income for the entire electricity sector. The MLE functioned like price cap regulation (see Chapter 4), but it was highly desegregated and had an explicit cost basis. The MLE established "standard costs" for all factors of production. These were used to produce tariffs that generated revenues for the distribution companies. The MLE mechanism then distributed the revenue to both generators and distributors according to their standard costs. The main characteristics of the MLE were the following:

- The cost recognition system was based on predetermined standard costs; the aim was to promote efficiency of electric service through cost minimization.
- The utilities were remunerated according to the predetermined standard costs, whatever their actual costs. Actual costs below standard costs meant profits for the utility; the opposite meant economic losses.
- A periodic cost review was used to determine the financial health and stability of the electric sector.

According to MLE regulation, total national revenues from electricity should match the national cost of service. This total cost was obtained by aggregating all recognized costs and dividing them by the expected demand. These costs included

- *Generation:* fixed costs of generation (amortization and remuneration of generation assets), fixed and variable operation and maintenance costs, and fuel costs
- *Distribution:* amortization of investment of distribution assets, operation costs, commercial management costs, structure costs, and financial expenses
- *Transmission:* all costs associated with the REE
- *Other expenses:* uranium stock, canceled nuclear projects (due to political decisions), etc.

The standard cost regulatory structure provided incentives for efficiency. Since the firms knew what their revenues would be under the MLE, they knew that cost reductions would increase profits. Also, there were incentives to (1) improve the availability of generating units, (2) extend the operating lives of assets, and (3) reduce distribution losses. All of these factors were incorporated into the standardization process. Some standard costs were adjusted annually in response to changes in price indices and interest rates. If the Ministry of Industry and Energy (which oversees the MLE) required extraordinary investments, those costs were incorporated into the tariff-making process. Environmental costs were also recovered in this way.

The MLE succeeded in improving efficiency. For example, real electricity prices declined at an average rate of 1.1% per year between 1988 and 1994. Average availability of the coal plants improved, and the use of power for auxiliary services at thermal plants declined. The financial performance of the firms, as measured by profitability, self-financing, and external debt, also improved (Kahn, 1995).

8.2.1.3. The 1994 Electricity Act (LOSEN). In the late 1980s, the government started to consider moving to a market-oriented competitive electricity sector. Motivations for this new orientation included the following:

- The organization of the industry as a vertically integrated natural monopoly was no longer necessary because of changes in generation technology; other countries, like Argentina and England and Wales, had successfully introduced competition.
- It was believed that a competitive generation sector would reduce costs and encourage innovation.
- Tariffs should reflect marginal costs, to provide the right incentives to consumers and investors.

The first stage in the recent restructuring process of the Spanish electrical sector was undertaken in 1994. The main items that characterized the sector were the following (Pérez-Arriaga, 1997, 1998):

- Electrical utilities were mostly private companies. Endesa Holding was expected to be privatized. The economic and financial health of the utilities was good.
- There had been a high level of concentration in generation and distribution (two utilities controlled 80% of the market).
- Installed generation capacity exceeded peak demand by a high margin, implying overcapacity.
- The regulated average electricity price was higher than the short- and long-term marginal price. The price was higher than expected under competition. Prices were higher than those in neighboring, and possibly competing, countries.
- A subsidy to the coal industry affected the economic operation of the generation system.

A new Electricity Law, Ley de Ordenación del Sistema Eléctrico Nacional (LOSEN), was approved in December 1994 under a socialist government. It was a first attempt to introduce competition into generation and to meet the expected requirements that the future European Directive on the Internal Energy Market would impose. Many arrangements foreseen in the LOSEN were not finally implemented. However, LOSEN did create an independent regulatory commission, Comisión Nacional del Sistema Eléctrico (CNSE).

8.2.1.4. The Electricity Protocol. In 1996, the Ministry of Industry decided to propose a more radical change than previously proposed by LOSEN: the introduction of competition into generation. At that time, the European Directive on the Internal Energy Market (96/92/CE) had been approved and a conservative political party was in control of the government. In December 1996, the Ministry of Industry and the utilities signed an agreement, the Electricity Protocol, the main points of which were:

- Setting the basis for the new competitive market in generation and establishing January 1, 1998 as the starting date
- A progressive reduction of regulated electricity tariffs over four years: at least 3% in 1998 (it was 3.63%), and 1% in each of the following years 1999, 2000, and 2001
- Recognition of the possibility of recovering stranded costs through the Costs of Transition to Competition (CTTC) as a fixed remuneration during a maximum period of 10 years and a maximum amount of 1,988,000 million Pta (US$13B)
- A timetable for customer choice: qualified customers were allowed to purchase electricity from any supplier

Retail competition would be phased-in over a 10-year period. As of January 1998, all customers with annual demand greater than or equal to 15 GWh, and railway companies, including metropolitan railways, could choose their supplier (i.e., they would be "qualified" customers). Table 8.13 presents the customer choice schedule approved in the Electricity Protocol and, later, in the new Electricity law.

8.2.2. The 1997 Electricity Law

The Electricity Law 54/1997 was approved in November 1997. This law lays the foundations for deregulation of the industry according to international experiences and goes beyond the guidelines of the European Directive. Besides opening generation and supply activities to competition, it modifies the sector's business structure, requiring legal unbundling of regulated and unregulated activities, and promoting new entry. These changes would be gradually phased-in during a transitional peri-

Table 8.13. Schedule for Customer Choice in the 1997 Electricity Law

Year	1998	2000	2002	2004	2007
Annual Consumption	15 GWh	9 GWh	5 GWh	1 GWh	All

Source: Electricity Law 54/1997; thirteenth transitional provision.

od, establishing a 10-year deadline for a complete liberalization (Unda, 1998). The basic principles of the law are the following.

- Less state intervention by rationalization of the energy policy constraints and by leaving system operation and planning to the electricity market (except transmission planning).
- Separation of activities. Regulated activities, such as transmission and distribution activities (including purchases by franchised regulated customers) are separated from nonregulated activities, such as generation and the provision of services to qualified customers. Generation and retail competition were introduced. Legal separation of regulated and unregulated activities was completed by December 31, 2000. This separation is restricted to the activities, not to ownership, so the same shareholder can have an interest in regulated and nonregulated activities.
- The design of a bulk power competitive market, including (1) competition in generation, started in January 1998, (2) freedom of entry, (3) a nonmandatory power pool managed by a MO, (4) free bilateral contracts (physical and financial) among agents, and (5) equal participating conditions for both the generation side and the demand side.
- Nondiscriminatory access to the network is guaranteed to all participants in the market.
- Transmission and distribution network businesses are considered as regulated natural monopolies with regulated transmission and distribution tariffs paid by all network users.
- Redefinition of the functions of the Regulatory Commission (Comisión Nacional del Sistema Eléctrico-CNSE) to achieve and promote competition and to supervise the transparency and objectivity of system operation.
- Creation of the MO (economic management of the standardized power transactions in the bulk power market) and of the SO (technical operation and supervision/control in bulk power system security). Both institutions should be independent, public or private. Also, any shareholder should not own more than 10% of equity capital, either directly or indirectly. The set of the wholesale market agents should not own more than 40% of equity capital.
- A calendar for customer choice established when and what size of customers would become qualified customers (see Table 8.13).

8.2.3. Further Regulations

The Electricity Law was amended through Royal Decrees. For example, Royal Decree 2017/97 (BOE, 1997b) established a general procedure for the settlement of regulated activities. Royal Decree 2019/97 (BOE, 1997c) set the rules for the wholesale electricity market. Royal Decree 2820/98 (BOE, 1998) accelerated the liberalization process set by the Electricity Law with the following schedule for qualified customers: January 1999, customers with annual consumption greater than 5 GWh; April 1999, 3 GWh; July 1999, 2 GWh; and October 1999, 1 GWh. Furthermore, Royal Decree 2066/99 (BOE, 1999) established that all customers connected to networks with voltages greater than 1 kV become qualified as of July 2000. Finally, Royal Decree 6/2000 (BOE, 2000a) set January 2003 as the date by which all customers would become qualified.

8.3. THE WHOLESALE ELECTRICITY MARKET

Competitive mechanisms now govern electricity generation. The main feature of this market is its institutionally flexible design. Transactions are conducted in two ways: either through an organized electricity pool (nonmandatory poolco model) with dispatch based on generation and demand bids or bilaterally between two parties. Figure 8.6 presents a general representation of the different types of transactions and the sequence of markets.

8.3.1. General Market Institutions

The economic and technical management of the system rests on two basic organizations:

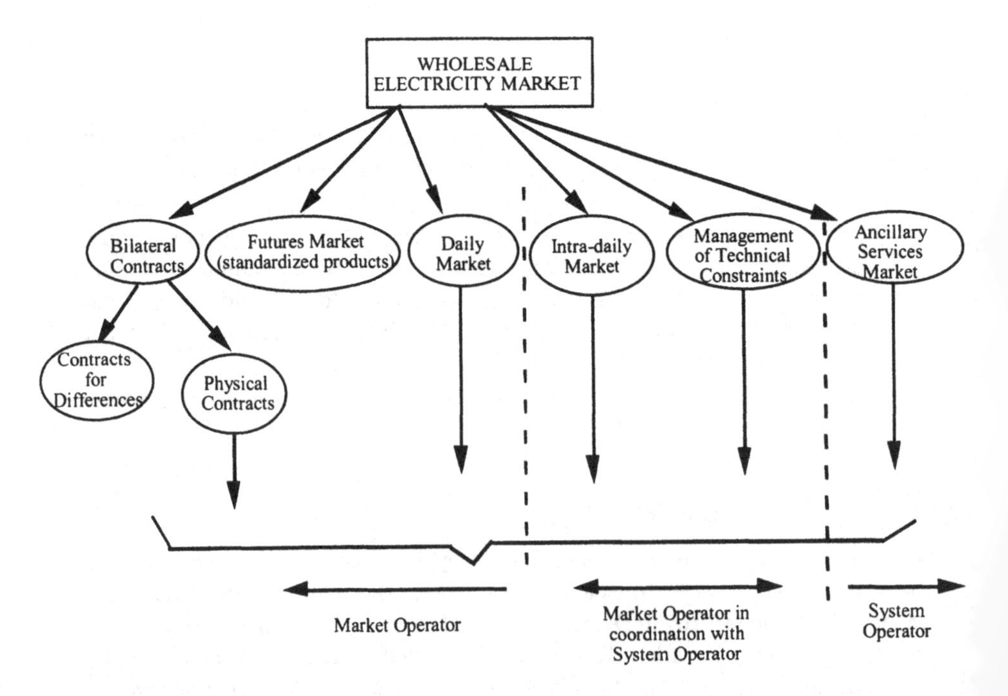

Figure 8.6. The Spanish wholesale electricity market. *Note:* A futures market does not exist in Spain. *Source:* Comision Nacional del Sistema Eléctrico, CNSE (1998).

1. The MO is in charge of the economic management of the system. Its primary functions are to receive and match the energy offers and bids and to carry out settlements. The MO is Compañía Operadora del Mercado Español de Electricidad SA (OMEL).
2. The SO is in charge of the technical management, i.e., the activities related to (1) the administration of the energy flows, taking into account exchanges with other interconnected systems and (2) the determination and allocation of transmission losses and the management of the ancillary services. The SO is Red Eléctrica de España SA (REE).

The National Commission of the Electricity System (CNSE), since 1999 integrated into the National Commission of Energy (CNE), the regulator is responsible for resolving conflicts arising from the economic or technical management of the system, as well as conflicts related to transmission or contracts for third party access to the transmission and distribution networks. The Market Agents Committee (Comité de Agentes del Mercado—CAM) represents all agents of the system in the generation market. CAM reviews the economic management and can propose measures to improve generation market operations.

8.3.2. Structure of the Wholesale Market

Royal Decree 2019/97 (BOE, 1997c) structures the generation markets as follows:

- The daily market
- The ancillary services market
- Physical bilateral contracts

Additionally, since April 1998, there are intraday markets, which manage the necessary adjustments to the daily program. There is also a procedure to manage transmission constraints. Each market has an independent price. Generators are remunerated based on their corresponding participation in each market at the resulting market clearing prices.

The MO centrally manages the wholesale markets. Market participants place forward bids to sell or buy energy. Each day there is a daily market (day-ahead) and five intradaily markets. Market participants are producers (generation, cogeneration, and renewable units), distribution companies (who buy energy for resale to regulated customers), suppliers or retailers (who buy energy to sell to qualified customers), qualified customers, and external agents (in neighboring systems). The MO matches selling and buying bids by using a matching algorithm. All transactions in these markets are firm.

8.3.3. The Daily Market

In the day-ahead market, each participant bids prices and quantities for the following day's 24 hours. Agents are allowed to present simple or complex sale offers to the MO. Simple sale offers can include up to 25 ranges of generation capacity for each generating unit, with a different price for each level. Complex sale offers contain the same information as the simple offer plus extra conditions that are considered by the matching algorithm to obtain a feasible schedule for generators, such as minimum value for cost recovery, ramp constraints, and must-run conditions.

The matching algorithm matches simple offers and independent generation with demand for each of the 24 hours, resulting in hourly marginal prices and dispatch of load and generation. The market price is set through the following mechanism.

- Once the period for offer presentations has expired, the MO matches the offers for each hour, in order from the cheapest one to the most expensive one, until the demand is met; it also takes into account the different constraints imposed by the complex offers. The matching algorithm determines a marginal price for each period. The marginal price corresponds to the sale offer of the last generation unit whose acceptance has been necessary to meet demand. The matching algorithm also provides (1) the energy committed by each agent and (2) the economic priority order for all the generation units that submitted offers (even for those that were excluded from the matching process).
- The MO communicates the matching result to the SO and to those agents who submitted offers in the daily market.

8.3.4. The Intraday Markets

These markets are used to account for deviations between the day-ahead market and what market participants want to do based on updated information on unit availability and demand deviations. There are five intradaily markets every day. Agents are allowed to present simple or complex sale offers to the MO. Simple offers can include up to five ranges of generation capacity (or five ranges of demand volume) for each generation or demand offer unit, with a different price for each range. In each intraday market, the MO matches generation and demand bids to establish the hourly program for each hour left until the programming day finishes.

8.3.5. Network Constraint Management Procedures

The SO applies a network constraint management procedure during the sequence of the day-ahead energy market operations and in real time. The SO, taking into account the quantities (generation and demand) that have been scheduled for every hour in the day-ahead energy market, makes a network analysis to evaluate possible congestion or voltage problems. If there are congestion or voltage problems, the SO will modify the results of the daily market, minimizing the cost of the deviations with the following:

- Forced-in generators (from constrained-on units) are paid the offer price that they submitted in the day-ahead market for the electricity generated in those scheduling periods in which those units are called upon to solve transmission constraints.

- Displaced units receive no compensation for their constrained off-generation.
- Ad hoc procedures can be defined for permanent constraints.

If constraints appear in real time, the SO can resort to emergency procedures. The additional cost incurred in removing all grid constraints is added to the cost of ancillary services. These costs are recovered through an uplift to the energy price and charged to the total demand in each hour.

8.3.6. The Ancillary Service Markets

The SO also manages a set of ancillary service markets with transmission congestion management procedures. Initially, secondary reserves (Automatic Generation Control, AGC) and tertiary reserves (spinning, nonspinning, and replacement) are being provided through secondary and tertiary reserve markets. The participants compete in quantity and price and receive the resulting marginal price for each of these independent markets. Other ancillary services, such as primary reserve (frequency response), reactive and voltage support, and system restoration, are provided, based on mandatory technical rules. It is intended that market mechanisms will be gradually introduced whenever possible to provide and remunerate these services.

8.3.7. Capacity Payments

Besides energy payments at market prices, generators also receive a capacity payment that depends on the annual performance. The capacity payment takes into account the generator's availability and its annual energy production. One reason for this is that in Spain demand bids cannot set the market price. So, the market price might not always reflect the value of capacity to customers. Since demand cannot set the price, an adjustment is needed to ensure that the short-term value of capacity is correctly perceived by consumers. However, the Royal Decree Law 6/2000 (BOE, 2000a) modified the high initial capacity payment, set at 1.3 Pta/kWh times gross demand, to 0.80 Pta/kWh.

8.3.8. Bilateral Trading

There are two types of bilateral trading: financial and physical. (There is also the possibility of a standardized futures market run by the MO.) Transactions subject to physical bilateral contracts need not go through the pool, although the market and the system operators must be notified about the amounts involved, so that the contracts can be taken into account in the corresponding schedule. Physical contracts do not have priority in the system with additional security conditions or privileges and they must pay their corresponding part of transmission and other regulatory costs. Financial contracts do not affect the system operation; the settlement of these contracts is made separately by the involved parties. The MO must be notified of these financial contracts.

8.3.9. International Exchanges and External Agents

Finally, the new organization of the wholesale energy market includes the liberalization of international electricity exchanges. To date, REE has handled these exchanges through long-term contracts with France and short-term operational exchanges with France, Portugal, and Morocco. The new participants will be free to accomplish international transactions, either through the power pool or by means of bilateral contracts. Despite this liberalization, the Ministry of Industry and Energy must authorize international agents to bid in the pool. Even so, according to the European Directive, the Ministry can only refuse permission if there is a lack of reciprocity (Unda, 1998 and CNSE, 1998).

8.4. TRANSMISSION ACCESS, PRICING, AND INVESTMENT

The Electricity Law guarantees all authorized agents, producers, distributors, suppliers, qualified customers, and external agents (third-party) access to transmission and distribution networks. In compliance with the provisions of the European Directive, the Law provides open access to the network on payment of the corresponding network charges approved by the Government. Only nondiscriminatory legal, technical, and economic rules can be applied.

8.4.1. Remuneration of Transmission Activities

In Spain, the proposed remuneration scheme for transmission activities is based on a revenue limitation formula applied during the first regulatory period from 1998 to 2001. The remuneration received by a transmission company j in the year t is calculated as (BOE, 1997b)

$$TR_{j,t} = TR_{j,t-1} \cdot [(1 + (IPC_t - 1))/100] + IIN_{j,t-1} \tag{8.1}$$

where

$TR_{j,t-1}$ is the remuneration received by the transmission company j in the previous year

IPC_t is the retail price index in the current year, in percentage terms

$IIN_{j,t-1}$ is the increment of the remuneration associated with the new transmission investments that company j put in operation in the previous year

The SO centrally coordinates transmission planning. After approval, the transmission company executes new network investments and the associated standard costs are taken into account in its remuneration.

8.4.2. Transmission Network Charges

Transmission network charges are embodied in the regulated full-service tariffs (i.e., tariffs for regulated customers) and in the access tariffs (for qualified customers who have exerted their rights). The term "network charges" might cause confusion because transmission tariffs are not separate from distribution tariffs in Spain. Qualified customers must pay access tariffs to use transmission and distribution lines. The "recognized transmission revenues" are collected through full-service and access tariffs from distributors by the regulator, which then allocates revenues between Red Eléctrica and other transmission owners, according to predefined revenue entitlements. Section 8.5 describes the allocation procedure.

The regulated full-service tariffs and access tariffs are set to cover transmission and distribution costs and other institutional and specific regulated costs. Voltage levels in six categories differentiate the tariffs:

1. Low voltage
2. 1 to 14 kV
3. 14 to 36 kV
4. 36 to 72.5 kV
5. 72.5 to 145 kV
6. more than 145 kV

Further, the tariffs have two separate components:

- A capacity term as a function of the requested demand (MW)
- An energy term as a function of the requested energy (MWh)

In medium-voltage and higher-voltage grids, access tariffs are time-differentiated by day and season.

8.4.3. Transmission Losses

In Spain, transmission losses are taken into account directly in the daily energy market. They are not considered as an explicit network charge. Generation and demand, taking into account transmission losses, are matched in the day-ahead energy market. Initially, under the regulatory transition period, transmission losses are charged (1) to qualified customers through standard loss coefficients and (2) to distribution companies that share the difference between actual and standard losses in proportion to their hourly demands (BOE, 1997c). In addition, a proposal to provide location signals related to losses to the new agents connected to the grid has been discussed and approved. Each new market participant connected to the Spanish transmission network (generator or load) will have an associated transmission loss participation factor (loss penalty factor) that depends on its marginal contribution to losses calculated at the point of connection. Transmission loss factors are escalated by the ratio of marginal to estimated losses, so the corresponding loss factors equal estimated losses. Generators in exporting areas will be remunerated for less energy than produced, whereas consumers in importing areas will pay for more energy than consumed. These loss factors are taken into account to match generation and demand in the central market model with a single pool price (OMEL, 1998a).

8.4.4. Investment and Planning

All grid users can promote construction and planning of new transmission facilities, but the SO must coordinate different proposals. All new proposed facilities should be considered when evaluating development plans for the network. Construction, operation, and maintenance of the new facilities will be accomplished through competitive bidding mechanisms. Authorized new investment should result in new allowed revenues for the transmission owner.

8.5. DISTRIBUTION NETWORK REGULATION AND RETAIL COMPETITION

The Electricity Law 54/1997 introduced retail competition by allowing specific customers (qualified customers) to buy energy directly from the wholesale energy market or through new retail companies specifically created for this purpose. Distribution and selling energy to still-regulated customers are considered regulated activities carried out by distribution companies. Accounting and legal separation are required between regulated distribution businesses and qualified customers who buy electricity at retail. Therefore, the supply of services to nonfranchised customers (qualified customers that have exerted their eligibility), is made under competition with other suppliers or retailers (in Spain, they are called "commercializers"). (The term "qualified customer" is equivalent to "eligible customer." The terms "nonregulated customer," "nontariff customer," and "free customer" can be used to denote the qualified customer who has used the eligibility right.)

8.5.1. Remuneration of Regulated Distribution Activities

The Electricity Law established that the method used to determine the remuneration of regulated distribution activities should be based on objective, nondiscriminatory, and transparent criteria. Also, the method should consider the geographic and market characteristics of the different distribution areas. BOE (1997b) set the remuneration of distribution companies through a revenue cap formula for the regulatory period 1998–2001. The remuneration that distribution company j will receive in year t, including distribution activities and energy supply to franchised customers, is calculated as (BOE, 1997b)

$$DR_{j,t} = DR_{j,t-1} \cdot [(1 + (IPC_t - X))/100] \cdot [1 + (\Delta DR_{j,t} \cdot Eff)] \qquad (8.2)$$

where

$DR_{j,t-1}$ is the remuneration received by the distribution company j in the previous year

IPC_t is the retail price index in the current year, in percentage terms

X is the productivity factor, set to 1% for all distribution companies

$\Delta DR_{j,t}$ is the annual per-unit increment in demand delivered by the company in the current year (if the demand decreases, then this value is equal to 0)

Eff is the efficiency factor that represents the per-unit increment in distribution revenues associated with a per-unit increment in delivered demand

Under the previous regulatory framework (MLE), recognized or standard distribution costs to remunerate distribution companies were computed as follows:

- High-voltage (U > 36 kV) installation costs were evaluated by considering standard costs associated with the type and number of installations the company put into service
- Medium- and low-voltage installation costs were evaluated by considering standard coefficients that multiply the annual energy flow supplied by each network

Under the new regulatory framework, the use of "starting-from-scratch" planning models for distribution networks has been proposed to set the initial revenue for each utility considering the geographic and market characteristics of its distribution areas (DISGRUP, 1998 and Román, Gómez, Muñoz, and Peco, 1999).

8.5.2. Distribution Losses

In addition, distribution companies have an explicit incentive to reduce losses in their networks. Standard loss coefficients are taken into account to calculate the standard network losses for distribution companies. These standard losses are charged to consumers through full-service and access tariffs. Table 8.14 shows the standard loss coefficients applied to the energy consumed by nonregulated customers connected to high-voltage networks. The standard loss coefficient applied to regulated customers connected to low-voltage networks is 13.7% (BOE, 1999).

Table 8.14. Percentage Energy Losses for Nonregulated Customers with Access Tariffs

Network Voltage Level	Period 1	Period 2	Period 3	Period 4	Period 5	Period 6
(1kV < U < 36kV)	6.8%	6.6%	6.5%	6.3%	6.3%	5.4%
(36kV < U < 72.5kV)	4.9%	4.7%	4.6%	4.4%	4.4%	3.8%
(72.5kV < U < 145kV)	3.4%	3.3%	3.2%	3.1%	3.1%	2.7%
(145kV < U)	1.8%	1.7%	1.7%	1.7%	1.7%	1.4%

Source: BOE (1999).

Note: The year is divided in six load periods.

Standard loss coefficients are calculated taking into account average estimated losses. Every distribution company must buy real losses from the energy wholesale market at the energy pool price. If the real losses are lower than the standard ones, then the company will make profits. Otherwise, it will sustain a loss.

8.5.3. Distribution Network Charges

Usually, aggregated network distribution costs are charged to final customers depending on the network voltage level where they are connected, despite their particular point of supply. A hierarchical structure of the network is assumed, so that customers connected by a high- or medium-voltage network pay a network charge that corresponds to their participation in the total costs of transmission, and upstream distribution network, according to (1) their requested demands, and (2) their peak responsibility factors.

Distribution network and transmission charges are embodied in the regulated full-service and access tariffs; see Section 8.4.2. These tariffs are classified according six different voltage levels, and they have two separate components: a capacity term and an energy term. In addition, they are time-differentiated by six periods during the year. Table 8.15 shows the range of variation of high-voltage access tariffs, depending on the period of the year.

8.5.4. Power Quality Regulation

Power quality regulation has been approved in Spain with specific regulation for transmission and distribution (BOE, 2000b). Most aspects of service quality are regulated: continuity of supply, voltage quality, and commercial services to regulated customers.

Distribution companies are responsible for technical aspects of the quality of service, including continuity of supply (i.e., interruptions) and voltage quality (i.e., voltage levels and voltage disturbances). Retailers selling energy to nonregulated customers can require the regulated technical quality from the distribution companies that supply their customers. Distribution companies must also control voltage disturbances by their customers. Therefore, customers must comply with limits of disturbance emissions set to guarantee voltage quality.

Table 8.15. High-Voltage Access Tariffs

Network Voltage Level	Capacity Term ($/kW-year)	Energy Term ($/MWh)
MV (1kV < U < 14kV)	12.7–2.1	2.2–0.6
MV (14kV < U < 36kV)	10.0–1.7	1.7–0.5
HV1 (36kV < U < 72,5kV)	9.2–1.5	1.6–0.4
HV2 (72,5kV < U < 145kV)	8.5–1.4	1.5–0.4
Transmission (145kV < U)	7.7–1.3	1.3–0.4

Source: BOE (1999).

Note: The range corresponds with time differentiation from Period 1 (highest value) to Period 6 (lowest value). Original data in Pta. Converted to U.S.$1 = 150 Pta.

Continuity of supply is measured and controlled in the different geographic supply areas of an electric company. The following reliability indices are calculated in each area:

- *Average System Interruption Duration Index* (TIEPI in Spanish) measures the total duration of interruptions. It is calculated as the average total duration of interruptions in medium voltage weighted by the affected installed capacity in medium voltage/low voltage transformers in a particular area. An 80 percentile TIEPI is the value of TIEPI met by 80% of the towns in the considered geographic area, for instance in a province.
- *Average System Interruption Frequency Index* (NIEPI in Spanish) is calculated as is TIEPI, but considers the number of interruptions, instead of duration.

Four areas are considered for control of continuity of supply: (1) urban areas—towns with more than 20,000 consumers; (2) semiurban areas—towns with more than 2000 consumers and less than 20,000; (3) rural concentrated areas—towns with more than 200 consumers and less than 2000; and (4) rural dispersed areas—towns with less than 200 consumers or areas with isolated consumers. Reliability indices are measured for each type of area in each province, and for each company. Measured indices should be below the specified quality limits proposed in Table 8.16.

If a distribution company does not comply with the required quality, it must propose a quality improvement plan that must be approved by the Regional Administration. If the plan is not carried out or delayed, the company could be penalized.

Continuity of supply levels for individual customers are also guaranteed. The total annual duration of interruptions and the total number of interruptions should be below specific limits, depending on the voltage supply level and the distribution area where the customer is connected. For instance, domestic customers connected in low-voltage networks should not have more than 6 hours (urban areas), 10 hours (semiurban), 15 hours (rural concentrated), or 20 hours (rural dispersed) of interruptions in a year. In case of noncompliance with these limits, the distribution company must compensate the affected customers with five times the average price of its annual consumption by the estimated energy not delivered in the hours of interruptions exceeding the limit.

Table 8.16. Proposed Reliability Index Limits for Continuity of Supply

Area Type	TIEPI (hours)	TIEPI 80% (hours)	NIEPI (# of interruptions)
Urban	2	3	4
Semiurban	4	6	6
Rural concentrated	8	12	10
Rural dispersed	12	18	15

Source: BOE (2000b).

Distribution companies will measure all aspects of voltage quality, as defined in European Norm 50160 (CENELEC, 1994). This information is sent to the Regional Administration and to the regulator. In addition, distribution companies are required to inform potential customers of expected voltage quality indices at their connection points.

Commercial services are also regulated. Distribution companies are required to meet specific quality levels regarding commercial services related to measurement, billing, maximum time for connection of new customers, quality of service information, etc. Companies should inform customers about best tariff options and energy use.

8.6. PARTICULAR ASPECTS OF THE REGULATORY PROCESS IN SPAIN

8.6.1. Estimated Stranded Costs

An adequate treatment of CTTC is a key issue when the introduction of competition is done in systems with private or investor-owned utilities, as in Spain or in the US. This problem has not appeared in countries where the liberalization process has been joined to the privatization process, as in England and Wales, Chile, Argentina, etc.

In the Electricity Protocol, the Ministry of Industry and the Spanish utilities agreed on a mechanism for recovering a limited amount of stranded costs during a transition period of 10 years. The stranded costs or CTTC were understood as the difference between the incomes that utilities would receive under the previous regulatory framework and the estimated incomes that they would receive under the new regulatory system based on competition.

The remuneration of the generation assets under the previous regulatory framework (MLE) was based on average generation costs calculated according to standard cost parameters. Because of new generation technologies, such as combined-cycle gas turbines (CCGTs) and international coal prices, the expected long-term marginal generation costs in Spain under a competitive framework would be lower than historic average costs. Therefore, the incomes of the generation utilities would decrease in comparison to the previous situation.

The process of computing the CTTC amount has been as follows (Arraiza, 1998):

- Computation of the present value of the income that the utility would receive under the previous regulatory framework (MLE)
- Computation of the present value of the expected incomes under competitive mechanisms based on marginal-cost pricing. A pool price of U.S.$40/MWh was assumed. This price included the capacity payment and the ancillary services costs.

- The difference between these two quantities was reduced by an efficiency factor of 32.5%.
- An amount of U.S.$1.7B has been added to the previous amount to subsidize the use of domestic coal in competitive conditions instead of imported coal.

The total resulting amount was U.S.$13B. Table 8.17 details this quantity.

8.6.2. The Stranded Costs: Methodology for Recovery

8.6.2.1. Initial Methodology (January 1998). All customers should pay the CTTC, as a regulated charge, without discrimination regarding regulated or nonregulated customers. The CTTC is charged through regulated full-service and access tariffs. Monies devoted to this compensation for the companies' stranded costs are the "Retribución Fija" (RF). However, the RF is a residual quantity, being paid from whatever is left after paying the different regulatory entitlements from the tariff revenue. These regulatory entitlements include payments for power purchased from the various markets (production, intraday, deviations, and reserve energy) for the tariff market (at the relevant wholesale market price), distribution remuneration, transmission remuneration, nuclear moratorium costs, MO and SO costs, and other costs. The residual is distributed to the companies as the RF payment for that year, according to the fixed percentage shares (see below for more detail on the settlement procedure to recover the regulated costs).

If the annual average pool price stands above the estimated pool price (US$40/MWh), then the annual amount of CTTC would be reduced by the difference between the actual pool price and the estimated pool price. So, the estimated value of US$40/MWh acts as a cap for the pool price for CTTC recovery. On the other hand, if the average pool price is below the estimated pool price, then the annual amount of recovered CTTC will be greater than expected and an acceleration

Table 8.17. Components of the CTTC Calculation in Spain

Concepts		Total (US$ billion)
Nuclear	7.9	
Thermal (coal and oil)	4.3	
Hydro and pumping	2.6	
New investments	1.2	
Other concepts	0.7	
Subtotal		16.7
Efficiency factor (32.5%)	–5.4	
Coal national subsidy	1.7	
CTTC Total		13.0

Note: Original data in Pta. Converted to U.S.$1 = 150 Pta.

of the CTTC recovery process would occur. Unless the principle of cost recovery is accepted and applied to tariffs in the future, there is no guarantee of full recovery of stranded costs.

8.6.2.2. Further Developments. In September 1998, the electricity companies and the Ministry of Industry and Energy agreed to some changes regarding the CTTC, as part of the revisions to the Electricity Protocol. These changes included the following.

- A write-off from the maximum CTTC entitlement, reducing it by approximately 85%.
- Securitization of about 75% of the CTTC entitlement; this amount will be recovered from 4.5% of the tariff billing over time.

An amendment to the Electricity Law was approved in November 1998 to allow these changes. However, after negotiations with the European Commission, the securitization proposal was never effectively implemented.

8.6.3. The General Settlement Procedure: Regulated Tariffs and Revenues

During the regulatory transition period in Spain, a settlement procedure for regulated activities in coexistence with nonregulated activities has been adopted. This procedure is presented in Figure 8.7. Regulated tariffs must compensate, apart from transmission and distribution revenues, part of the incomes of the agents who participate in the competitive generation market, such as fixed costs of some generators (stranded costs) or additional costs for special generation (renewable or cogeneration) (BOE, 1997b).

The regulated incomes received by each distribution company are collected in three ways:

1. *Regulated full-service tariffs* paid by final franchised customers, including energy and network services
2. *Regulated access tariffs* paid directly by free customers or by retailers for their free customers
3. *Connection fees* and *metering equipment fees* paid by end-use customers

The tariffs are calculated by taking into account that the total incomes should cover the total electricity sector costs. Table 8.18 presents this cost estimation.

Some of the regulated full-service tariffs and access tariffs are devoted to support the institutional and specific system costs, such as system operator, market operator, and regulator costs, and compensation for nuclear investments that were made in power plants that were canceled, etc. Usually, the percentages applied to full-service tariffs are different from those applied to access tariffs. This is to ensure that the same amount is being recovered whether the customer is free or not.

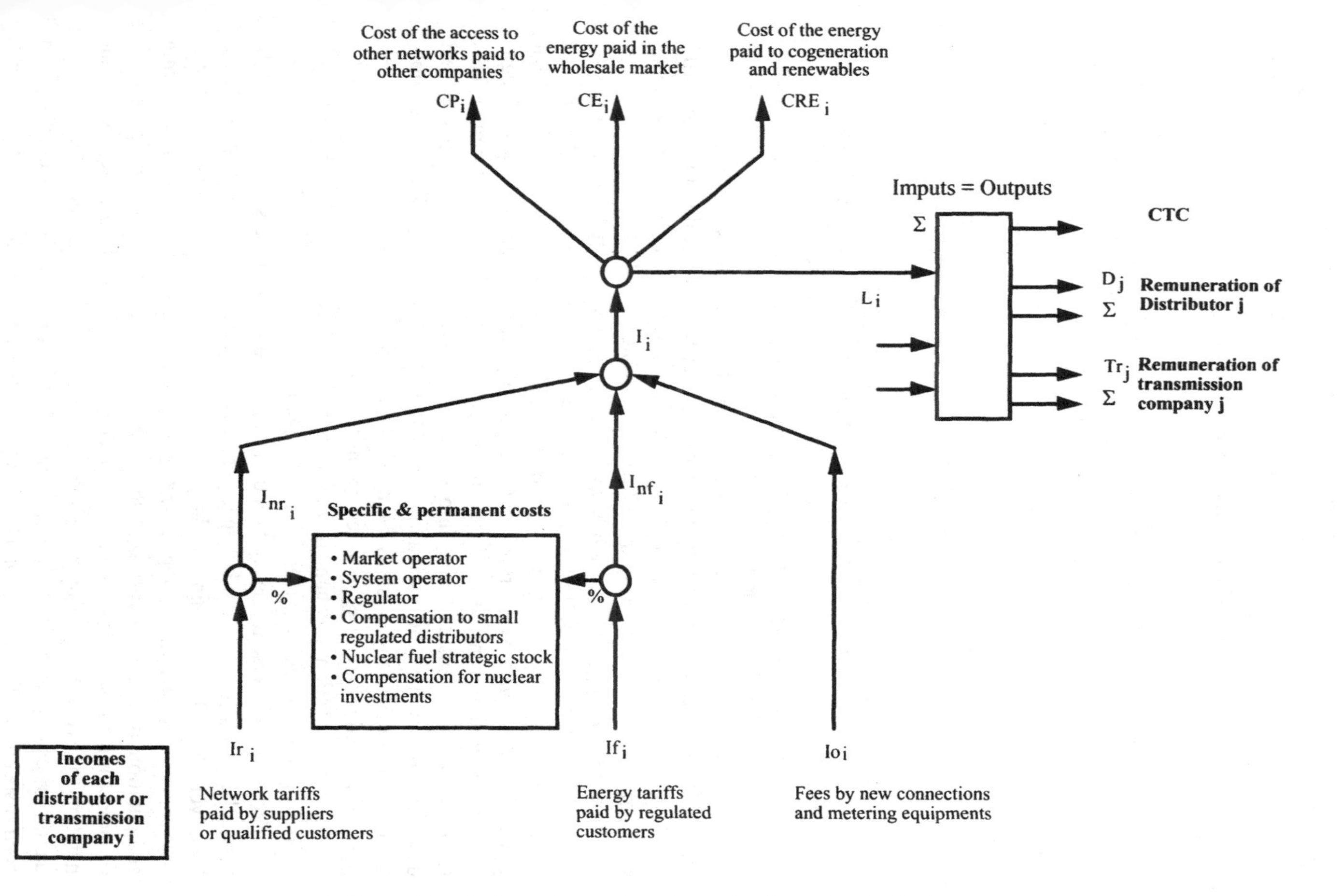

Figure 8.7. Settlement procedure of regulated activities in Spain.

Table 8.18. Costs Included in the Estimation of Regulated Tariffs in 1998

Cost Concepts	1998
Generation costs (Pool energy $30.6/MWh, Ancillary services $1/MWh, Capacity payments $8.7/MWh, Renewables and other)	54.8%
Competition transition costs (CTTCs)	10.7%
Institutional costs: SO, MO, CNSE, other	1.8%
Transmission costs (REE plus transmission owners)	4.8%
Network distribution plus supply to regulated customers costs	23.5%
Specific costs for generation diversification: nuclear moratorium and others	4.4%
TOTAL COSTS	U.S.$ 12.9 B

Source: UNESA (1998).
Note: Original data in Pta. Converted to U.S.$1 = 150 Pta.

Each distribution company must pay for the energy (pass-through) that has been purchased in the wholesale market to other generators or to special generators. This is the energy that the distribution company has bought to supply electricity to its regulated customers. The distribution company must also pay the corresponding access tariffs to other distribution companies.

Because of the CTTC recovery procedure and because access tariffs are not differentiated by geographical areas, the company net income would not necessarily equal the company regulated revenues set by the regulator. Therefore, the compensation mechanism shown in Figure 8.7 has been implemented.

Under this compensation mechanism, every distribution company puts its net income in a common fund. The sum of the total net incomes from all distribution companies is equal to the sum of the total regulated revenues plus the CTTC that generation companies would receive. Every distribution company receives its regulated remuneration from the fund, according to the revenue limitation formula. This remuneration is calculated taking into account the estimation of the companies real distribution costs, i.e., the particular characteristics of the region where the company distributes electricity, such as the number of customers and the customer densities in rural and urban areas supplied, as discussed above.

The compensation mechanism decouples direct payments made by network users from companies' remunerations. The economic efficiency of this practice has been criticized because it does not send the right economic signals to customers. In particular, the differentiation between rural and urban customers is not considered. However, it has been recognized that, at least during the transition period, it provides beneficial regulatory effects. First, it is appropriate for the standardization and computation of the distribution costs, because the regulator, by using uniform (nondiscriminatory), transparent, and objective criteria, establishes the base remuneration for every distribution company in the country. There are no different evaluation criteria for each distribution company. This problem usually occurs when each distribution company proposes its own network tariffs based on its own cost estimation and evaluation. Second, it contributes to the open access principle, because the network tariffs will not discriminate between customers as a function of the company that provides the service. Third, it has been established that the structure of access tariffs paid by free customers should be the same as the structure of full-service tariffs paid by regulated customers.

In this way, the following relationship is met:

$$\textit{Access Tariff} = \textit{Full-Service Tariff} - \textit{Wholesale Energy Price} \qquad (8.3)$$

such that the following conditions are met.

- There is no discrimination between regulated and nonregulated customers; all customers contribute to the regulated system costs.
- Coherency is achieved between economic signals, i.e., network and energy prices.
- There is a unified settlement procedure for regulated tariffs, i.e., access and full-service tariffs.
- The tariff system takes into account voltage levels and time of use, which provides efficiency by giving economic signals to the users.

CHAPTER 9

THE ARGENTINE POWER SECTOR

9.1. GENERAL DESCRIPTION OF THE ARGENTINE POWER SYSTEM

In the late 1980s, Argentina faced a chronic lack of investment, high growth in electricity demand (nearly 7% per year), and frequent power outages. In response to this challenge, Argentina introduced a competitive market in 1992, with the Electricity Law #24.065. Generation, transmission, and distribution were separated. Most of the electricity companies were privatized. In 1994, important reforms to the bidding system were enacted. Argentina's electricity sector at the wholesale level became one of the most open and competitive in the world. Unfortunately, today, Argentina is facing a severe national economic and financial crisis whose consequences on the electricity sector are difficult to predict.

9.1.1. Generation and Current Structure of the Industry

In 1992, the government carried out a vertical and horizontal disintegration of all federal electricity companies, except those with nuclear plants and the binational hydroelectric projects. Then, the resulting business units were privatized. From 1995, most of the provincial companies were also vertically separated and privatized. Since 1992, generation has been organized as a competitive market with independent generation companies selling their production in the wholesale market, Mercado Electrico Mayorista (MEM), and by private contracts with other market participants. The administration and operation of MEM is done by the Compañía Administradora del Mercado Eléctrico Mayorista SA (CAMMESA), which acts both as system and market operator. Section 9.3 describes the wholesale market in more detail.

Currently, there are more than 40 private thermal and hydroelectric generating companies operating in MEM. The largest are the thermal plants, Central Puerto (1009 MW), Central Costanera (1260 MW), and ESEBA (1049 MW), and the hydroelectric plants, Chocón (1320 MW), Alicurá (1000 MW), and P. del Aguila (1400 MW). In addition there are two state-owned binational hydroelectric power generators: Yacyretá (1500 MW, Phases I & II, owned jointly with Paraguay), and Salto Grande (540 MW, owned jointly with Uruguay). Finally, there are two state-owned nuclear plants (1108 MW); a third nuclear plant of 750 MW has been under construction since 1981.

In the liberalized Argentine market, the private sector has been eager to fund investment in generation capacity. Statistics on generation capacity and private ownership are illustrative:

- 2.0 GW of new gas-fired capacity were added to the system since 1991 and at least as much is under construction.
- From 1991 to 1997, total installed capacity increased by 31% from 14.6 GW to 19.1 GW.
- The number of generating companies increased from 14 to 45 from 1991 to 1997.
- 40 of these 45 generating companies are privately owned.

(These data do not include provincial and cooperative generators that do not belong to the MEM.) Annual electricity production was 74,500 GWh in 1999 and the installed capacity was about 19,920 MW. The stable regulatory and political environment made generation investment in Argentina attractive.

In 1999, 34.3% of the energy was provided by hydroelectric units, 56.7% was by fossil-fuel-fired units, and 8.5% by nuclear power plants. The rest came from imports. Hydroelectric units are mostly located far from the Buenos Aires area, where almost 60% of the load is concentrated. They are connected through radial transmission lines. Gas is imported by pipeline from the Comahue and the Northern Argentine regions to the Buenos Aires area, where most gas-fired generation units are located.

9.1.2. Transmission

Transmission activity is regulated. Transmission companies are required to provide third-party access to all market participants and collect network charges for providing transmission services. Transmission companies are prohibited from generating or selling electricity. The law precludes any generator, distributor, large customer, or any firm (that controls or is controlled by any of them) being the owner of or the controlling shareholder of a transmission firm.

The Compañía de Transporte de Energía Eléctrica en Alta Tension SA (TRANSENER) is the major independent and regulated transmission company. It owns and operates the 500 kV interconnected network and some of the 220 kV lines. This interconnected system (SADI in Spanish) covers almost 90% of Argentina. TRANSENER's lines link the main generation areas with the main demand center of Gran Buenos Aires. The 500 kV network is a radial network with a low level of interconnection. When any of these major lines fail, the supply to Buenos Aires can be seriously disrupted, limiting the available energy. A critical transmission corridor is between the Comahue area and Buenos Aires. Because of Comahue generation additions, this corridor has reached full capacity. The construction of a so-called "fourth line" to expand transmission capacity linking the north and south of the country has been concluded.

Apart from TRANSENER, there are other transmission companies with high-voltage transmission assets: six private regional transmission concessionaires and 14 Technical Transmission Function Providers (PAFTTs). Table 9.1 shows the most important transmission companies.

Currently, there are international links with Uruguay and Brazil, and exports to Chile are expected. Recent construction includes 1000 MW of interconnection capacity with Brazil and another 1000 MW are under construction. It is expected that Argentinian sales will release pressure in the domestic market, where many generators are competing for market share. Most of Yacyretá's (hydroelectric) output will be exported (even with 7% annual load growth), so that new entry of thermal generators will be encouraged. Transmission expansions are developed under competition by privately owned, independent transmission owners; see Section 9.4.4 for more details.

9.1.3. Distribution

Distribution is also regulated. It involves the transmission of electricity from the supply points in the bulk power system, owned by TRANSENER and the regional transmission companies, to final customers. Distribution companies operate as geographic monopolies under licenses, with the obligation to provide service to all regulated consumers within their specific region.

There are four types of distribution companies. First, at the national level, there are three private national distribution companies (EDENOR, EDESUR, and EDELAP) that operate in the Gran Buenos Aires area. The national regulatory body, Ente Nacional Regulador de la Electricidad (ENRE), regulates them. Second, there are provincial private distribution companies that are regulated by provincial regulatory bodies. Third, there are provincial electric utilities that have not been privatized; some of them are vertically integrated. They are not explicitly regulated. Fourth, there are many cooperatives under the jurisdiction of provincial or munici-

Table 9.1. Transmission Companies in the Argentine Interconnected System

Transmission Companies	Voltage Levels (kV)	Lines (km)	Transformers (MVA)
TRANSENER	500/220	7,450	6,300
TRANSNOA	132	2,464	956
TRANSPA	300/132	2,217	1,248
DISTROCUYO	220/132	1,245	1,025
TRANSNEA	330/132	836	601
TRANSCOMAHUE	300	829	387

Source: CAMMESA: www.cammesa.com.

pality authorities. In 1997, private companies distributed almost 50% of the energy. Table 9.2 gives output and number of customers as of 1999.

9.1.4. Consumption

Electricity demand has had an average annual growth rate of 6.2% from 1990 to 1998. In 1999, the industrial sector accounted for 40% of the total demand; residential and commercial users accounted for 31% and 17%, respectively. Other uses accounted for the rest. Most of the consumption, 55%, is concentrated in the Buenos Aires area (city and Gran Buenos Aires). The annual electricity demand invoiced to final customers in 1999 was 64,489 GWh (see 1999 electrical sector data at http://energia.mecon.gov.ar).

The Electricity Law recognizes a class of large customers, consisting of industrial and other users with substantial electric supply needs, that can buy energy from the spot market or through contracts with generators, outside the regulated tariffs. Large users include consumers whose peak demand is equal to or higher than 30 kW. Large users are divided into three categories:

1. Major large users (GUMA in Spanish) larger than 1 MW with annual consumption larger than 4380 MWh who are required to sign contracts for at least 50% of their purchases and take the remainder of their energy at the MEM spot price.
2. Minor large users (GUME in Spanish) between 0.1 and 1 MW with triple tariff measurement equipment who must contract 100% of their load.
3. Particular large users (GUPA in Spanish) between 0.03 and 0.1 MW with simple tariff measurement equipment.

Further, GUMEs could also be customers with a peak demand between 0.03 and 0.1 MW with triple tariff measurement equipment, or with a peak demand between 1 and 2 MW, contracting 100% of their load. There are plans for expanding retail competition to all customers. In 1999, there were 409 GUMAs, 1545 GUMEs, and 29 GUPAs.

Table 9.2. Distribution Companies and Zones: Energy and Customer Supplied in 1999

Distribution Companies and Zones	Invoiced Energy (GWh)	Customers (thousands)
EDENOR	12,881	2,237
EDESUR	12,324	2,105
Buenos Aires	8,053	1,601
Santa Fe	5,722	979
Cordoba	3,918	902
EDELAP	1,933	284
Other provinces	19,656	3,204

Source: Secretaría de Energía y Minería: www.energia.mecon.gov.ar.

9.1.5. Electricity Tariffs

From the beginning of the privatization process (September 1992), average tariffs for typical customer classes have been kept almost constant (Guidi, 1997). For instance, low-demand (up to 10 kW) residential customer tariffs (T1-R2) have been near U.S.$65/MWh. General use customer tariffs (T1-G2) have had a tariff near U.S.$110/MWh. Tariffs for high-demand customers (above 50 kW) connected to medium-voltage (MV) networks (T3-MV) were stabilized at U.S.$60/MWh.

9.1.6. Economic and Energy Indices

(Source: www.eia.doe.gov/emeu/cabs/)

- Population (2001E): 37.4 million
- Size: 2.8 million square kilometers
- Gross Domestic Product (GDP, 2001E): U.S.$273 billion
- Real GDP Growth rate (2001E): –1.5%
- Inflation rate (2001E): –0.7%
- Oil production (2000): 816,100 barrels per day (bbl/day)
- Net oil exports (2000E): 312,100 bbl/day
- Natural gas production (1999E): 1.22 trillion cubic feet (Tcf)
- Natural gas consumption (1999E): 1.19 (Tcf)
- Coal Production (1999E): 370,000 short tons
- Coal consumption (1998E): 1.35 million short tons
- Energy consumption per capita (1999E): 74.2 MBtu (versus 355.8 MBtu in the United States)

9.2. THE REGULATORY FRAMEWORK

9.2.1. Background

Until 1992 in Argentina, as in other Latin American countries, the electric sector was mainly composed of government-owned enterprises. Federal and state companies owned most generation. Table 9.3 shows the electricity generation by ownership in 1987.

Tariffs were set by Cabinet decision, although there was a governmental agency in charge of electricity policy. In Argentina, the Ministry of Public Works and Services, through the office of the Deputy Secretary of Energy, had authority to control the electricity sector's planning, licensing, tariffs, and development. The Ministry of Economics controlled tariffs and investment programs. In addition, public electricity companies' investment decisions were substantially controlled by the Federal Government. In particular, long-term investments were curtailed during periods of macroeconomic adjustment and periods of political instability.

Table 9.3. Electricity Generation by Ownership in 1987

Company Owner	Generation of 48,084 GWh
Federal	89.5%
Provincial	10.3%
Cooperatives and others	0.2%

Source: Spiller and Martorell (1996).

Public ownership, nontransparent regulatory systems, and direct intervention by the administration on pricing and investment decisions influenced pricing policies (Spiller and Martorell, 1996). Some effects were

1. Average tariffs below long-run average costs
2. Cross-subsidies and discrimination among final customers
3. Uniform prices across regions

Consequently, at the end of the 1980s, Argentina had chronic electricity shortages, the operation of the electricity generation system was highly inefficient, and large generation investments were canceled or delayed, giving priority to smaller and lower-capital investment plants.

At that time, some tariff categories were cross-subsidizing other tariff categories, which sometimes did not cover energy costs. Also, the government subsidized this system every year. There were no quality standards and no relationship between tariffs and quality of supply. Thermal generation unavailability levels of 50% (due to lack of maintenance) and dry hydrological conditions during years 1988 and 1989 caused severe energy restrictions, producing long and frequent supply interruptions. Under this situation, the Federal Government decided to introduce major changes by restructuring and privatizing the electricity industry.

9.2.2. The New Electricity Law

Law 24,065 was issued in 1992 (Electricity Law, 1992), changing the previous 1960 Law 15,336 and establishing the new regulatory framework. The main objectives of the Law were (Bastos and Abdala, 1996)

- To protect the users' rights
- To promote competition in generation and supply markets
- To promote the operation of transmission and distribution under adequate conditions of reliability, open access, and nondiscrimination
- To ensure fair and reasonable tariffs
- To promote new long-term investment to ensure supply and competition wherever possible

The main results of this Law were (Bastos and Abdala, 1996)

- The separation of generation, transmission, and distribution
- The creation of a wholesale electricity market based on marginal prices and competition among generators (low levels of ownership concentration were allowed; a maximum of 10% was specified)
- The transmission was regulated and transmission owners were not allowed to participate either in generation or in distribution
- The creation of CAMMESA, the System Operator (SO) in charge of the economic and technical system operation
- Distribution was carried out within each service territory under the terms of a distribution concession (obligation to supply) that recognized monopoly conditions in the corresponding area
- The creation of ENRE as the regulatory commission

The successful privatization process that was put in place in Argentina (Pérez-Arriaga, 1994)

- Promoted the entry of private capital in a well-defined and competitive framework
- Promoted economic efficiency, improved investment, and plant availability
- Ensured cost-reflective tariffs based on marginal costs
- Limited the government's involvement to the regulatory roles of promoting efficiency and protecting customers

The privatization took place by splitting generation, transmission, and distribution into separate companies and selling the majority of their shares by open competitive tender. See Section 9.2.4 for a more detailed description of the privatization process.

9.2.3. Regulatory Authorities

The Secretary of Energy, who works within the jurisdiction of the Ministry of Economy, Public Works, and Utilities, has overall responsibility and regulatory power over the electricity industry. Currently, the Secretary grants and controls electricity sector concessions at the national level through the National Directorates for Coordination and Regulation of Prices and Rates for Electricity Planning. The Secretary also receives assistance from the Federal Board of Electric Energy, which has representatives from each province. This is an advisory body that coordinates policies for the sector.

ENRE was created by the Electricity Law with the functions of a regulatory commission. ENRE is an independent body with legal capacity to pursue the stated objectives of the Law. ENRE reports to the Secretary of Energy. ENRE has five independent members selected by the Government. ENRE enjoys considerable autonomy and its members cannot be easily removed. The main functions of ENRE are

- To enforce Law 24,065,
- To issue standards of safety, quality of service, and environmental protection
- To prevent anticompetitive, monopolistic, or discriminatory behavior
- To establish the basis for tariff determination
- To establish the criteria and conditions for awarding concessions
- To control the sector activities and apply penalties for noncompliance

The main objectives regarding quality of service, tariff, and obligation of supply are centered on the rights of final users. This new model separates the regulatory functions (in the hands of the Federal Government) from the service itself, which is provided by private firms. Transmission and distribution are "public services," whereas generation is only defined as a "general interest" activity. The first two activities are regulated by ENRE, whereas generation is subject to market rules (except safety and environmental concerns that are supervised by ENRE). Distribution companies have the obligation to supply and the government assumes no responsibility for guaranteeing the security of supply. Penalties for failing to meet this obligation are imposed whatever the reason for the failure.

CAMMESA is a nonprofit organization responsible for scheduling and dispatching generating units to meet electricity demand. CAMMESA also coordinates the payment and settlement of MEM transactions. CAMMESA's shareholders are (1) the Secretary of Energy and associations representing (2) generators (AGEERA), (3) transmitters (ATEERA), (4) distributors (ADEERA), and (5) large users (GUMAs), each holding 20% of its equity. The government appoints its chairman and vice-chairman, and retains specific powers of veto. The Chief Executive Officer is nominated by the five shareholders.

9.2.4. The Privatization Process in Argentina

The assets of the former state-owned utilities SEGBA, AyEE, and Hidronor were split into many small companies. Most of their shares have been sold. The privatization is almost complete; see Tables 9.4–9.6. The policy of the two binational hydroelectric plants has still to be decided. Once the restructuring process is complete, private capital will be included in nuclear power plant ownership.

9.3. THE WHOLESALE ELECTRICITY MARKET

Competition in generation was introduced with the creation of a wholesale spot market (MEM). Offers to sell energy made by generators provide the basis for both the economic dispatch and the setting of energy spot prices in the MEM. Distribu-

Table 9.4. The Privatization of Generators

Name	Privatization Date
Central Puerto SA	April 1992
Central Costanera SA	May 1992
Central Pedro de Mendoza	October 1992
Central Dock Sud	October 1992
Central Alto Valle	August 1992
Central Guemes	September 1992
Central Sorrento	January 1993
Central San Nicolas	April 1993
Central Termicas del Noreste Argentina SA	March 1993
Central Termicas del Noreste	March 1993
Central Termicas Patagunicas SA	November 1993
Central Termicas del Litoral SA	July 1994
Hidroelectrica Diamante SA	September 1994
Hidroelectrica Rio Hondo SA	December 1994
Hidroelectrica Ameghino SA	October 1994
Centrales Termicas Mendoza SA	October 1994
Hidroelectrica Alicura SA	August 1993
Hidroelectrica El Chocon SA	August 1993
Hidroelectrica Cerros Colorados	August 1993
Hidroelectrica Piedra del Aguila SA	December 1993
Hidroelectrica Futaleufa SA	June 1995

Source: Ministry of Economy and Public Works and Services: www.mecon.ar.

Table 9.5. The Privatization of Transmission Companies

Name	Privatization Date
TRANSENER	July 1993
TRANSNOA SA	January 1994
TRANSPA SA	January 1994
TRANSNEA SA	October 1994
DistroCuyo	December 1994

Source: Ministry of Economy and Public Works and Services: www.mecon.ar.

Table 9.6. The Privatization of Distribution Companies

Name	Privatization Date
EDENOR SA	August 1992
EDESUR SA	August 1992
EDELAP (La Plata)	November 1992
Empresa electrica de Rio Negro	August 1996
Compañía General de Electricidad Jujuy (CGE)	November 1996
Empresa Distribuidora de Energía Norte (EDEN)	April 1997
Empresa Distribuidora de Energía Sur (EDES)	April 1997
Empresa Distribuidora de Energía Atlantic (EDEA)	April 1997

Source: Ministry of Economy and Public Works and Services: www.mecon.ar.

tors and large users (with direct access to MEM) purchase energy at the contractual, seasonal, or spot price.

The wholesale generation market is organized as follows:

- A single market for the bulk trading of electricity (the pool)
- Hourly market prices for electricity sold by generators, but average seasonal prices for electricity purchased by distributors (a mechanism for balancing seasonal differences)
- A central role for the national dispatch entity (CAMMESA) in maintaining system security through central scheduling and dispatch of generating units
- A central settlement system managed by CAMMESA for the pool transactions between the pool members

9.3.1. Market Participants

The main participant categories in the MEM are the following:

1. Generators subject to scheduling and dispatch rules that have financial contracts with distributors, retailers, or large users; these contracts do not affect the order of economic dispatch
2. Transmitters responsible for the bulk-power transmission system
3. Distributors responsible for distributing energy to final customers through medium-voltage (MV) and low-voltage (LV) networks
4. Large users that buy energy directly from the MEM or through retailers
5. Retailers that buy energy produced by generators and sell energy consumed by large customers

9.3.2. Energy Market and Economic Dispatch

Economic dispatch is designed to ensure the most efficient production of electricity. Initially, MEM procedures required thermal generators to bid their audited variable cost of production, which included their fuel costs. (Since November 1995, generators have been allowed to bid prices; upper limits to the price bids are set.) In Argentina, CAMMESA sets generators' offer prices based on the seasonal fuel prices. Hydroelectric generators also submit their units' water values to CAMMESA. The hourly spot price resulting from the economic dispatch is generally set by the most expensive thermal generator running.

The hourly energy spot price is differentiated according to nodal factors that account for the marginal effect of transmission network losses. Nodal factors are calculated for each system transmission bus, taking as a reference the system load center in Ezeiza, close to Buenos Aires. Nodal factors vary according to season and time of day (the peak is from h18 to h23, the valley is from h23 to h6, and the shoulder hours are from h6 to h18). Nodal factors reflect the expected marginal losses associated with an incremental energy transaction from a transmission bus to the load center. In exporting areas, nodal factors will be lower than one. In importing areas, they will be higher than one. The marginal energy price at a given node (price received by the generator or paid by the consumer) is computed as the product of the corresponding nodal factor times the pool's marginal energy price. Nodal factors are also modified in case of transmission congestion; see Section 9.4. (In other restructuring experiences, as in California or Norway, nodal or zonal prices are defined only to consider the effect of congestion, whereas transmission losses are considered through "loss factors"; see Chapters 6 and 7.)

9.3.3. Capacity Payments

In the Argentine system, a capacity payment is added to the energy spot price to promote expansion capacity and efficient operation. This capacity price reflects the generator's contribution to the system reliability margin during nonvalley hours of weekdays. The capacity payments are diverse, and depend on several conditions; see Pérez-Arriaga (1994) and CAMMESA (2000). Each MW of dispatched capacity receives this supplement during nonvalley hours. Each MW of capacity that is available and scheduled, but not dispatched during nonvalley hours, also receives this supplement. (Note that the capacity payment has been reconsidered, particularly for generation plants included in dispatch. Given conditions of excess supply, generators compete to earn this payment by reducing their offer energy prices.)

Each MW of base load plant (fossil-fueled and nuclear units, but not hydroelectric or gas turbines) also receives the capacity payment for its estimated capacity factor during a hypothetical extra-dry year. If expected output in a dry year exceeds actual output in the current year, plants receive a supplemental payment for the difference. The intention is to encourage investment in generation capacity that contributes to the long-term reliability of the system, taking into account the uncertainty associated with hydro conditions. Payment in 1999 was set at U.S.$10/MWh of expected deliveries in an extra-dry year (less receipts for other capacity payments).

Capacity payments to generators are made for plant capacity, net of bilateral contract commitments. Capacity payments are affected by "adaptation" factors. These factors adjust the value of capacity in relation to transmission reliability. CAMMESA simulates the effects of network contingencies and reliability of the network link with the system load center. Generators sending power to Buenos Aires from a remote location receive less than generators inside the Buenos Aires area. Discounts in generator revenues from the application of nodal and adaptation factors (as an implicit charge for using the transmission system) are discussed in Section 9.4.1.

9.3.4. Cold Reserve and Ancillary Services

In addition, generators are paid for providing operating reserves (cold reserve) and ancillary services (frequency regulation associated with spinning reserves, voltage regulation, reactive power control, and start-up services). CAMMESA requires cold reserves to ensure the secure operation of the system. Every week, it organizes an auction in which peaking generators (gas turbines) bid prices, quantities, and response times to provide this type of reserve. All the scheduled reserves are paid the same marginal price, equal to the highest reserve bid price accepted. CAMMESA allocates the cost of these services to customers of MEM through additional charges.

9.3.5. Generator Revenues

As stated above, generators (apart from the nodal energy price revenue obtained from the wholesale energy market) can obtain additional capacity and ancillary services payments. The capacity payment is intended to cover (1) their capacity costs for running, (2) being scheduled in peak hours, or (3) even if they do not run, their forecast output in extra-dry years. The regulations state that a generator should receive the larger of either the reliability component (see 9.3.6) or the capacity payment, but not both.

9.3.6. Scheduling, Dispatch, and Settlement

Scheduling and dispatch of units is under the responsibility and control of CAMMESA. Maintenance coordination, hydro scheduling, and the computation of the "value" of water are made through medium-term seasonal studies with a three-year scope and a scheduled six-month period. CAMMESA uses hydrothermal coordination models to set the seasonal price paid by distributors for energy purchases from the pool. This seasonal price is a fixed price during a six-month period. This seasonal price of energy is increased by an additional reliability component that includes provision for loss of load probability in peak demand situations. The value of loss of load is fixed by the government and is currently U.S.$1500/MWh. There are different procedures to review and update the price if computational assumptions change significantly (CAMMESA, 2000).

CAMMESA carries out unit commitment scheduling weekly, taking into account information provided by generators. It prioritizes the units for dispatch. Finally, it controls and supervises daily generation dispatch and real-time operation, and calculates the corresponding hourly spot prices. CAMMESA acts as an agent for all participants in MEM to settle all market transactions and additional charges and payments.

9.3.7. Bilateral Contracts

The contract market in Argentina is based on financial contracts among market agents intended for hedging against price volatility and to secure supply. (There is no futures market in Argentina.) They are financial instruments that do not entitle the holder to physical delivery of electricity, although they do relate to a particular amount of electricity. The actual dispatch of generating units does not consider contracts among generators and distributors or large customers. Consequently, the generator can be dispatched to provide more or less energy to the pool despite its contractual commitments. Under these circumstances, the generator will be obligated to buy or sell the excess energy in the pool at the spot price.

Bilateral contracts must define injection and offtake nodes so that transmission loss factors and constraint costs can be determined. The details of bilateral contracts are publicly available. Only privately owned generators are allowed to sign contracts directly with customers or distributors.

On the one hand, the regulated sector does not impose financial risk on distribution companies from pool price uncertainty or volatility. Distribution companies purchase energy at seasonal nodal prices that are the same ones used to set the regulated tariffs. In this sense, the seasonal price is like an automatic contract-for-differences for distribution companies. On the other hand, distribution companies that are unable to supply their customers are subject to penalties according to their concessions. Penalties from failures to meet supply are the only risks that require hedging. In theory, these risks can be covered through long-term contracts with generators, since priority is given to those distributors with contracts for supply shortages. However, supply priority is effective only if, at the moment of energy deficit, (1) the generator is generating and connected to the MEM and (2) there are no transmission constraints isolating the buyer. Distribution companies so far have generally preferred to bear the risk of penalties caused by supply shortages. Excess generating capacity in recent years means that buyers are not facing a high risk of being unable to meet their demand requirements. It is not surprising, therefore, that only a little more than half of all energy purchases by distributors were made under contract (as of May 1997).

When they were privatized, distribution companies signed initial contracts with generators at above-market prices, to remunerate the sunk costs of generation. The most important of these contracts ran out in the year 2000. GUMAs are required to contract for at least 50% of their purchases at market prices and GUMEs must contract for 100%. Distribution companies have long-term contracts for more than 60% of their estimated demand.

In 1993, only 30% of the total amount of pool purchases were made through contracts and the rest were negotiated at the pool prices, mainly at seasonal rates. The percentage of purchases under contract increased to 60% in 1995. The average, nominal contract price in the market has decreased from U.S.$40/MWh in1994, to U.S.$32/MWh in 1996, to U.S.$29/MWh in 2000 (www.cammesa.com).

9.4. TRANSMISSION ACCESS, PRICING, AND INVESTMENTS

Transmission regulation applies to the SADI. Transmission companies are regulated by ENRE under the jurisdiction of the Secretary of Energy.

9.4.1. Transmission Charges

Transmission users pay several transmission charges to cover the operation and maintenance costs of the transmission system. These charges sum the regulated revenues of the transmission companies. Annual transmission charges include the following.

- A *connection charge* is based on capacity (for a generator) or maximum coincident demand (for a large user). This charge remunerates transmission assets for their connection to the transmission network. Values of U.S.$10/hour per connection at 500 kV, U.S.$9/hour per connection at 220 kV, and U.S.$8/hour per connection at 132 kV have been set. This was fixed at a maximum amount of U.S.$9M/year for the first 5-year regulatory period (Marmolejo and Williams, 1995).
- A *fixed capacity charge* is based on a share of the operation and maintenance network costs calculated at each node according the estimated use of the system capacity. The use of transmission lines is calculated with a load-flow model. It was fixed at a maximum amount of U.S.$30M per year to be recovered during the first 5-year regulatory period. After this first regulatory period, connection and capacity charges were reduced annually by an efficiency coefficient set by ENRE. A ceiling of 1% (after inflation) has been set for this coefficient, or 5% over the entire 5-year period (Marmolejo and Williams, 1995).
- There are two *variable charges*. A *reliability charge* (or *variable capacity charge*) is implicitly recovered as the difference between the capacity payments paid by distributors and those received by generators when using "adaptation factors." A *variable energy charge*, reflecting marginal losses, is implicitly recovered as the difference between the energy payments paid by distributors and those received by generators when using the "nodal factors."
- A *complementary charge* is based on estimated participation shares in each line. For existing network facilities, the allocation of the complementary charge to customers is based on the use of the network at annual peak demand. This complementary charge is a balancing charge equal to the difference between the actual amounts paid for variable charges and the expected transmission charges. The expected values are projections at five-year intervals taking into account the total cost that should be recovered by transmission companies. It was fixed at U.S.$55M per year during the first 5-year regulatory period.

Table 9.7 details the percentage of the total TRANSENER's revenues recovered by each type of transmission charge.

9.4.2. Penalties for Unavailability of the Transmission Assets

Transmission companies are subject to economic penalties or incentives associated with nonavailability of either transmission capacity or connections. If the provided level of availability is higher than a specified standard, companies receive bonuses,

Table 9.7. TRANSENER Revenues July 1993-June 1994

Charges	Revenues
Connection	9%
Fixed capacity	33%
Variable + complementary	58%
TOTAL	U.S.$ 91.2 million

Source: Marmolejo and Williams (1995).

otherwise they are penalized. This mechanism applies to all unforeseen outages. Scheduled outages bear a reduced penalty of only 10% of the specified rate for forced outages. The incentive/penalty mechanism accounts for (1) the duration of the outages, (2) the number of them, (3) the added charges that the outage will impose in the economic dispatch of generation or potential load curtailment, and (4) the type of line that becomes unavailable. Transmission companies collect an initial amount from user charges and they suffer a penalty for each failure or unavailability for which they are responsible. If the total penalty paid by a transmission company exceeds a specified limit, the Secretary of Energy can remove the transmission license.

9.4.3. Transmission Concessions

TRANSENER was granted the exclusive provision of high-voltage transmission public service in November 1992. The concession agreement was granted for 95 years. This concession period is divided into management periods of 15 years (the first period) and 10 years (subsequent periods). At the end of each management period, ENRE will call for bids for the controlling stake. If another offer exceeds TRANSENER's offer, the company would receive the offered amount, changing the ownership of the controlling stake. The government has the right to grant new concessions for subtransmission network operation and investment.

9.4.4. Transmission Expansion

The expansion of the transmission system is a critical issue if electricity demand continues to grow. TRANSENER does not have a direct responsibility for expanding the existing transmission system. When lines are constrained at their maximum capacity, the locational energy price yields a "congestion rent" equal to the difference between the energy prices (less incremental losses) at each end of the line, multiplied by line flows (in MWh). These funds are collected corridor by corridor and applied to a compensation fund (SALEX funds). This is used to fund new lines. New lines can be built in two ways (NERA, 1998):

1. *Expansion by private contract.* Users interested in building a new line form their own coalition and internally agree on how to assign annual amortized

costs among themselves. SALEX funds are not used to subsidize the new line. A cost/benefit analysis is not required. ENRE could veto the construction proposal.

2. *Expansion by public auction.* Users can suggest the transmission expansion when they represent 30% of the anticipated usage. CAMMESA calculates costs (canon fees) to potential users. SALEX funds are used to decrease the cost of the line, up to a limit of 70%. Users charged with paying at least 30% of the canon could form their own coalition to block the line. ENRE must show that the transmission expansion is cost effective for the whole system without considering the part coming from SALEX funds.

For example, for the Yacyretá hydroelectric project connection, the first 500 kV, 200 km line was constructed by Yacilec SA, an independent transmission company that receives a monthly canon fee of U.S.$2.38M. The construction of the second Yacyretá line was awarded to another consortium that will receive a monthly fee of U.S.$2.4M for 10 years. In both cases, the main line user is the Yacyretá generator. Similar expansion mechanisms have been used to build interconnection lines with Brazil.

Several drawbacks of the regulatory system for expansion of new transmission have been pointed out (NERA, 1998):

- Generators might have perverse incentives in both directions; they might not be willing to pay for economic lines, while having an incentive to over commit funds to expansion.
- The use of SALEX funds might encourage uneconomic construction.
- The combination of these factors could encourage uneconomic location decisions.

Complementary regulatory solutions have been discussed based on the introduction of congestion transmission rights and new ways of proposing and developing transmission installations by independent transmission owners.

9.5. DISTRIBUTION REGULATION

Argentina is governed as a federation of 23 Provinces and the Federal Capital of Buenos Aires. Each province has it own government, parliament, and courts. Each province has responsibility for the public services of its own territory. There is a federal electricity supply area that is responsibility of the Federal Government and includes the city of Buenos Aires and neighbor towns in the province of Buenos Aires. Before 1992, this area was supplied by the vertically integrated company, Servicios Eléctricos del Gran Buenos Aires SA (SEGBA SA), owned by the Federal Government. Distribution was divided into three areas with three licensed companies: EDENOR SA, EDESUR SA, and EDELAP SA. The first two companies serve 2,100,000 users each, and the third serves 250,000 users. Licenses were awarded in September 1992 for EDENOR and EDESUR, and in December 1992 for EDELAP.

Consortia bidding for these licenses required involvement of a distribution operator with enough experience and technical capacity to perform adequately. This operator was required to have a participation of at least 20% in the enterprise making the offer. If two operators were involved, they should have a joint participation of at least 25%.

The valuation of each of the privatized areas was based on the projection of expected revenues and efficient management. Privatization took place through the sale of 51% of the shares in public auction. Another 39%, still owned by the state, was sold in the stock market. The company's staff acquired 10%. The licenses were given for 95 years. Table 9.8 shows the total amount paid for the licenses.

Gran Buenos Aires was the first distribution area to be privatized. The privatization process has continued for distribution companies in other provinces. Each province is establishing its own criteria, since they have autonomy to do so, and all of them are complying with the National Law 24.065.

9.5.1. Distribution Concessions

The distribution company must supply any load within its service territory: distribution supply is compulsory under the terms of the distribution concession. Quality of service norms that are explicitly set in distributors' concessions must be met, and insufficient generation cannot be grounds for nonresponsibility during supply interruptions. Distribution services include network investment, operation, and retail supply. Large customers (over 30 kW) are allowed to become qualified customers participating in the wholesale electricity market.

Distribution cost evaluations set distribution charges at the beginning of each regulatory period. The first regulatory period was set for 10 years. After that, distribution charges will be reviewed every 5 years. The resulting tariffs and revenues are also updated according to the evolution of an external inflation index.

Price cap regulation is applied to set maximum prices for each consumer class. As in similar price control schemes, the distribution company can increase its profit by increasing its efficiency or reducing its cost; a rate of return is not guaranteed. The strong incentives for maximizing benefits through minimizing

Table 9.8. Prices Paid for the Distribution Companies

Company	U.S.$ million
EDENOR SA	428
EDESUR SA	511
EDELAP SA	139

Source: Ministry of Economy and Public Works and Services: www.mecon.ar.

costs, such as investment and maintenance costs, could cause a reduction of service quality. Therefore, penalties for supply interruptions apply to distribution companies (MEOSP, 1992). These penalties are paid as compensation to affected users.

The 95-year distribution concession is divided into management periods. The initial management period was set for 15 years, followed by consecutive 10-year periods (Marmolejo and Williams, 1995). At the end of each period it is possible to transfer ownership of the company. Then, ENRE will invite tenders for the next period license. The current distributor has the option of keeping the license or being paid the highest bid (other than its bid). This innovative approach is intended to eliminate complaints about distribution charges, and to introduce an element of competition into the assignment of distribution licenses.

Within each tariff period, distribution charges are updated every 6 months using a combined index of U.S. Consumer Price and Producer Price Indices. Also, every 3 months, the tariff component representing the energy price is updated according to the seasonal price estimates made by CAMMESA.

9.5.2. Evaluation of Distribution Costs

This section outlines the method for determining distribution costs, when distribution in Gran Buenos Aires (GBA) was privatized, and how these costs were assigned to different tariff categories. The details of this methodology have not been made public. Hence, this section has been written considering the available information at the time when conditions for the distribution licenses for GBA were announced (Guidi, 1997).

The following elements are considered as distribution costs according to the present regulatory distribution framework:

- Necessary investments for expansion and replacement of distribution networks
- Operating and maintenance cost of distribution equipment and installations as a function of the Average Incremental Cost (AIC); see EDF-DISTRELEC (1989)
- Metering, billing, and general commercial activities
- A return on equity

The AIC calculation is based on the Minimum Cost Expansion Plan, considering investment, operating costs, losses, and cost of energy not supplied. SEGBA SA developed the plan to aid privatization. The following is a reasonable representation of the AIC and related calculations for the distribution system depicted in Figure 9.1.

Distribution costs (Dc_j) for each voltage level network (j = HV, MV, and LV networks) are estimated as

$$Dc_j = (CRF_j + UOP_j) \cdot AIC_j \tag{9.1}$$

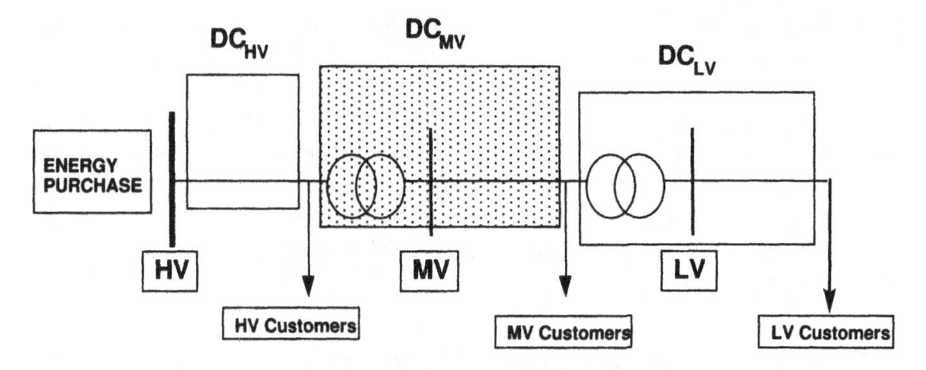

Figure 9.1. Network costs. *Source:* Guidi (1997).

where

CRF_j is the capital recovery factor at the selected discount rate

UOP_j is the annualized operating cost per unit of installed network capacity

AIC_j is the required level of network investment per unit of demand increase (measured in U.S.$/kW)

Equation 9.2 shows the calculation of AIC_j for each voltage level:

$$AIC_j = \text{NPV}(I_j)/\text{NPV}(\Delta P_j) \tag{9.2}$$

where

AIC_j is the average incremental cost for voltage level j

$\text{NPV}(I_j)$ is the net present value of minimum expansion (investment) cost for voltage level j, at the selected discount rate calculated over a 10-year period

$\text{NPV}(\Delta P_j)$ is the net present value of annual demand increments at voltage level j calculated during a 10-year period

Besides investment and operating costs, costs associated with commercial activities are also evaluated. The following customer-related activities are considered: (1) metering, (2) billing, and (3) general customer commercial activities. Regulated retail sales were not separated from distribution sales, because only large consumers can directly buy energy in MEM. For example, these activities were valued at the following charges for each type of customer:

- U.S.$1 per month for residential users
- U.S.$2/MWh for public lighting
- U.S.$0.25/kW per month for medium-size customers, between 10 and 50 kW

9.5.3. Regulated Tariff Customer Categories

Three important groups of customers are considered according to the requested demand. Each category has different subcategories according to their price elasticity.

Table 9.9 shows the regulated customer categories. For "low-demand" customers, only energy is considered. The electricity bill is based on a bimonthly fixed charge and an energy variable charge without any time block discrimination. An additional category division is made for residential and general users, taking into account energy consumption. For "medium-demand" customers, peak demand and energy are metered without any time block discrimination. For "high-demand" customers, tariffs depend on (1) power demand in peak and off-peak hours, (2) active energy divided in three time blocks (peak, shoulder, and valley hours), and (3) the consumption of reactive energy.

9.5.4. Cost Allocation in Regulated Tariffs—An Example of a User Tariff

The current tariff system is based on tariff formulas proposed for each category of customer. Wholesale seasonal capacity, energy, and transmission and distribution costs are allocated to each regulated customer following equity and economic efficiency principles:

- Wholesale seasonal capacity and energy prices, considering the corresponding "adaptation" and "nodal" factors, are charged to customer tariffs. Seasonal energy prices are calculated for three time blocks: peak, shoulder, and valley hours.
- Network distribution costs are distributed into day peak and night peak costs according to load curve studies at the different voltage levels. Cumulated network costs per kW supplied are computed for day and night peak hours for HV, MV, and LV customers. Each tariff category has a distribution net-

Table 9.9. Regulated Customer Categories

Category	Subcategory
Low demand (T1) (up to 10kW)	Residential: Up to 300kWh/bimonthly (R1) Above 300kWh/bimonthly (R2) General use: Up to 1600kWh/bimonthly (G1) From 1600 to 4000 kWh/bimonthly (G2) Above 4000kWh/bimonthly (G3) Lighting
Medium demand (T2) (from 10 to 50kW)	No subcategories
High demand (T3) (above 50kW)	Low voltage (LV) Medium voltage (MV) High voltage (HV)

Source: MEOSP (1992).

work cost component that represents its contribution to the total network costs. Commercial distribution costs are allocated among the different customer categories.

- Wholesale seasonal capacity prices and network distribution costs are closely related to the power demand in each time block (they are expressed in U.S.$/kW-month in day and night peak hours). Energy prices are expressed as U.S.$/kWh, and commercial costs are divided by the number of customers (U.S.$ per user month).

In the following, the tariff formulas for users that belong to the "low-demand" residential category are presented. The tariff formula consists of a bimonthly fixed charge and a variable energy charge. (In Equation 9.3a, the quantity in brackets is multiplied by 2 because it is a *bimonthly* fixed charge.) See Table 9.10 for definitions of the variables.

Table 9.10. Parameters of the Tariff Equation (9.3) for Residential Customers (R1)

Symbol	Definition	Values R1
Ppot	Wholesale capacity price	
Pep	Wholesale peak energy price	
Per	Wholesale shoulder energy price	
Pev	Wholesale valley energy price	
KrpL	LV power loss factor	1.143
KerL	LV energy loss factor	1.128
Kep (°/1)	Percent of purchased energy during the peak hours	0.27
Ker (°/1)	Percent of purchased energy during shoulder hours	0.63
Kev (°/1)	Percent of purchased energy during the valley hours	0.10
Kpp (kW)	A factor representing the incidence of the user capacity in the total capacity price	0.215
Dc (*dp*;LV)	LV network costs assigned to daytime peak hours in U.S.$/kW-month	4.81
Dc (*np*;LV)	LV network costs assigned to night peak hours in U.S.$/kW-month	5.30
KRdp (kW)	Responsibility factor of the user demand in the network installation development assigned to daytime peak hours	0.9*0.110
KRnp (kW)	Responsibility factor of the user demand in the network installation development assigned to night peak hours	0.9*0.215
Kfv	Allocation factor of the network costs between fixed and variable costs	0.75
GC (U.S.$-month)	Commercial costs per user	1.0
Kgc	Allocation factor of the commercial costs between fixed and variable costs	0.30
Ce (kWh)	Monthly energy consumption of the typical user of this category	27

Source: MEOSP (1992).

Fixed charge:

$$Fc = \{Ppot \cdot KrpL \cdot Kpp + [Dc(dp;LV) \cdot KRdp + Dc(np;LV) \cdot KRnp] \cdot Kfv + GC \cdot Kgc\} \cdot 2 \quad (9.3a)$$

Variable charge:

$$Vc = (Pep \cdot Kep + Per \cdot Ker + Pev \cdot Kev) \cdot KerL + [Dc(dp;LV) \cdot KRdp + Dc(np;LV) \cdot KRnp] \cdot (1-Kfv)/Ce + GC \cdot (1-Kgc)/Ce \quad (9.3b)$$

Tariff formula parameters are updated every 6 months to consider U.S. Consumer and Producer Price Index variations and the variation of economic indices.

9.6. PARTICULAR ASPECTS OF THE REGULATORY PROCESS IN ARGENTINA

9.6.1. Regulation of Power Quality after Privatization of Distribution

In Argentina, to ensure that the new regulation would not reduce investment and therefore reduce the quality of supply, the regulator focused on supply quality. In fact, it is one of the most complete regulations under the new framework of restructuring and competition. Some important aspects of this regulation are the following (MEOSP, 1992):

- *Distribution companies are responsible for the aggregated quality of the electricity product.* Reliability of generation, transmission, and distribution systems affects the supply quality to final customers. This induced distribution companies to participate actively in the proposals for reinforcements of the transmission grid and in new generation projects.
- *Three aspects of quality are regulated:* continuity of supply, voltage quality, and commercial services.
- *Penalties in case of noncompliance with minimum specified quality levels* are imposed. Distribution companies will compensate customers for quality levels below the guaranteed ones. Penalties for interruptions are based on an estimate of the Energy Not Served (ENS) cost experienced by affected customers.
- *Individual customer control:* Quality is controlled for each customer. To achieve this individual control, several intermediate control stages have been gradually introduced.

9.6.1.1. Quality Control. Three stages of quality control were included in distribution licenses: preliminary stage, first stage, and second stage. In the last one, individual customer quality control is implemented.

Preliminary stage: This stage was the first year after privatization. During this stage, there were no penalties. It was considered an adjustment stage for the next stages. ENRE and distribution companies prepared control mechanisms and made standardized reliability measurements.

First stage. This stage lasted for 3 years. Reliability system indices were measured at HV/MV transformers, at headers of MV distribution feeders, and at MV/LV transformers. Interruptions were classified according to the location of the failure, i.e., whether the failure occurred in the generation, transmission, or distribution segment. Different maximum limits were specified for each 6-month control period; see Table 9.11. The reliability system indices controlled were

1. The average number of interruptions in each MV/LV transformer (FMIT)
2. The total duration of unavailability of MV/LV transformers (TTIT)
3. The average number of interruptions weighted by the installed MV/LV capacity affected (FMIK)
4. The total unavailability duration of the installed MV/LV capacity (TTIK)

The limits for the last control period of the first stage are presented in Table 9.11.

Second stage (the definitive stage). The aim of this stage is to control the quality for each customer. All aspects of power quality are controlled: the number and duration of interruptions, and voltage disturbances, such as flicker, voltage fluctuations, harmonics, etc. The maximum limits are different for each voltage level (HV, MV, and LV) customer. The individual quality records for each customer are obtained from the database of customers' connections to the network and the registered interruptions. In this sense, no quality control equipment is installed for each customer connection. Voltage disturbances are controlled by measurement of a random sample of customers. Commercial services are controlled by setting maximum limits to different quality indices, such as

- The maximum time needed to connect a new customer to the grid (between 5 and 30 days)
- The maximum amount of estimated, instead of registered, consumption (< 8% of energy)
- The maximum time for resolution of complaints (within 10 days)

Table 9.12 presents the maximum limits related to interruptions.

Table 9.11. Limits for System Reliability Indices at the End of the First Stage

Distribution Failures	Generation and Transmission Failures
FMIT: 2.2 events/semester	FMIT: 2 events/semester
TTIT: 7.8 hours/semester	TTIT: 6 hours/semester
FMIK: 1.4 events/semester	FMIK: 2 events/semester
TTIK: 4.6 hours/semester	TTIK: 6 hours/semester

Source: MEOSP (1992).

Table 9.12. Limits for Customer Interruptions in the Second Control Stage

	Number of Interruptions per Semester	Total Interruption Duration per Semester
HV customer	3	2 hours
MV customer	4	3 hours
LV customer	6	6–10 hours

Source: MEOSP (1992).

If these limits are exceeded, distribution companies must compensate the affected customers. Compensation is calculated as a unitary ENS cost per kWh interrupted. The kWh interrupted is derived from registered or standardized customer demand curves.

9.6.1.2. Cost of Energy Not Served. The Energy Not Served (ENS) cost per kWh proxies how much customers would be willing to pay to have an adequate level of quality. In economic equilibrium, the customer would be indifferent between being interrupted or receiving the ENS value of being interrupted. In Argentina, this ENS cost has been set at U.S.$2/kWh. Studies identified ENS costs varying between U.S.$1 and $3/kWh depending on different consumer classes (ENEL–DISTRELEC, 1989).

If penalty values were equal to ENS costs, there would be an economic equilibrium for average customers and companies with an adequate quality of service. Under the implemented power quality regulation, distribution companies are expected to invest in an "economic" level of quality at which the marginal expected benefits of paying no penalties would be equal to the marginal investment costs for additional quality improvements. Figure 9.2 illustrates this.

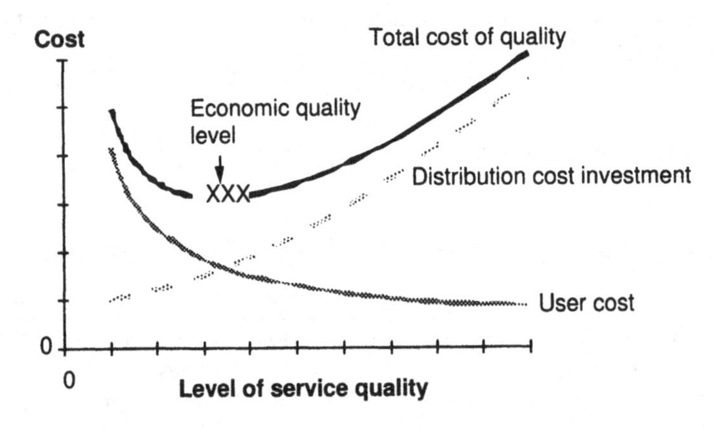

Figure 9.2. Cost versus quality. *Source:* Guidi (1997).

9.6.1.3. Results. Quality levels have improved significantly since quality regulation has been implemented. At the beginning of the process, quality levels were low, resulting in high compensation to customers. Most of the compensation (76%) was because of interruptions. Bad quality in commercial services was responsible for only 5% of the total compensation. Quality levels have gradually improved. Today, most of the companies have succeeded in attaining the required quality levels.

GLOSSARY

The section number in parentheses is the principle reference in the text.

Allowed rate of return (s) The rate of return on a firm's assets set by the regulatory authority. (4.2)

Ancillary services Technical services, such as operating reserves and voltage control, necessary to support a reliable interconnected transmission system; also known as interconnected operation services. (5.2.5)

Annuity (A) A uniform amount to be received or paid during a fixed number of periods. (3.2.1)

Asset An object or claim that is expected to provide benefits to its owner. Assets appear on the balance sheet. (3.2.3)

Asymmetric information A situation in which parties to a transaction or interaction possess different information. Its occurrence can lead to market failure or to the failure of regulation to achieve optimal outcomes. (4.3.4)

Average Cost (AC) Total cost divided by the quantity produced, equal to average fixed cost (AFC) plus average variable cost (AVC). (2.2.2)

Avoided costs Costs avoided by the regulated electric utility when distributing power from a nonutility generator. (4.1.1)

Balance sheet A statement of a firm's financial position, listing the firm's assets, liabilities, and equities, such that assets are equal to liabilities plus equities. (2.2.1)

Bilateral contracts Contracts between an individual buyer and an individual seller, generally outside a centralized market. (5.2.3)

Bonds A form of debt that can be secured with a specific asset or with the earning power of the borrower; see "Debt." (3.4.1)

Capacity payments Payments to generators to provide electricity generating capacity. Capacity payments are introduced to offset the lack of revenue or to encourage capacity investment. (5.3.1.1)

Capital Asset Pricing Model (CAPM) A model that shows that the risk premium on the purchase price of a capital asset depends on the correlation between the asset's return and the return on all capital assets. (Exercise 3.3)

Capital Recovery Factor (CRF) A rate that yields a uniform payment over a fixed number of periods such that the present value of these payments is equal to the present value of a capital asset. Also used in determining the levelized capital cost. (3.2.1)

Competition Transition Charge (CTC) A charge imposed by the regulator (e.g., in California) on customers to compensate electric utilities for losses due to the transition from a regulated to a deregulated market; see "Stranded costs." (6.6)

Costs of Transition to Competition (CTTC) A charge imposed by the regulator (e.g., in Spain) on customers to compensate electric utilities for losses due to the transition from a regulated to a deregulated market; see "Stranded costs." (8.2)

Congestion charges Charges imposed in network (e.g., transmission) pricing to the users of a congested line or interface equal to the cost saving of having an additional unit of transfer capacity. (5.3.3)

Consumer Price Index (CPI) A cost-of-living index, used to measure inflation, calculated as the cost of purchasing a set of consumer goods and services in comparison to the cost of the same set in a base period. (2.1.2)

Consumer surplus (*CS*) The difference between the total benefit derived by consumers of a good or service and the total amount paid to producers for the good or service. (2.4)

Contract for Differences (CfD) A financial contract in which the seller agrees to sell a specified quantity of electricity at a specified price during a specified period; the seller pays the difference between the market price and the contracted price. (5.2.2)

Correlation (CORR) A measure of the linear association between two variables, equal to the covariance divided by the product of the standard deviations. It varies between +1 for a perfect positive association and –1 for a perfect negative association. (Exercise 3.1)

Cost-of-service regulation (COS) A form of rate regulation in which the regulator compensates a firm for the cost of providing services to the customers; see "Rate-of-return regulation." (4.2)

Covariance (COV) A measure of the statistical association between two random variables. (Exercise 3.1)

Cross-price elasticity The percentage change in quantity demanded for one good or service in response to a one percent change in the price of another good or service. (2.1.3)

Cross-subsidization The subsidization of one good with the revenues from another, including the subsidization of one customer class (e.g., rural customers) with revenues from another (e.g., urban customers). (4.4.1)

Customer choice The ability of end-users to choose their supplier; also known as retail choice. (5.4)

Deadweight loss The sum of lost consumer and producer surplus. It represents the loss to the economy (society) of a market failure. (2.4)

Debt An obligation to repay a borrowed amount (of money) usually at a specified rate of interest for a specified time. Bonds are one form of debt. (3.1)

Demand (*D*) A *function* relating the amount consumers wish to purchase at each price. This must be distinguished from "demand" in electrical engineering, which refers to the instantaneous capacity (in MW) required by consumers (load). (2.1.2)

Demand elasticity (E_d) The percentage change in the quantity consumers demand in response to a one percent change in the price. (2.1.3)

Demand Side Management (DSM) Programs to reduce electricity consumption, for example, base load or peak load. (4.3.3)

Depreciation The decline of the value of an asset, for example, through use; usually accounted for as a cost of production in both tax and regulatory accounts. (3.2.3)

Discounting The recognition that money in hand today has a different value than the same amount one year ago or one year from now, reflecting the "time value of money." (3.2)

Dividends Payments to the firm's equity holders (owners). (3.4.1)

Economies of scale The percentage change in cost in response to a one percent change in output. If costs increase by less (more) than one percent with a one percent increase in output, there are *positive (negative) economies of scale.* If costs increase by one percent with a one percent increase in output, there are *constant economies of scale.* (2.2.3)

Efficiency, technical For a given technology, the greatest possible output for a set of inputs. (2.3.2)

Efficiency, economic For a given technology, the minimum opportunity cost to produce output. (2.3.2)

Energy Not Supplied or **Energy Not Served** (ENS) Electrical energy (in MWh) not supplied due to outage or supply interruptions. (9.6.2.1)

Equity Can refer to either (1) fairness or (2) equity capital. Equity capital in the form of equities, or shares, represents a share in the ownership of the firm. Payments to equity holders occur after payments to debt holders.

Expected return, $E(r)$ The mean return, equal to the sum of all possible returns weighted by their probabilities. (3.4)

Expected utility, $E(U)$ For an individual, the sum of all possible values (cardinal utilities) weighted by their probabilities; see "Utility function." (Exercise 3.2)

Externality When the production or consumption of some good or service affects the production or consumption of another good or service. Generally arises because the effect is not associated with a price and no market develops to facilitate exchange or regulation. (2.4)

Firm Transmission Rights (FTR) A contractual right that entitles the holder to receive a portion of the congestion charges or fees collected by the system operator. (5.3.3)

Fixed Cost (FC) Costs that do not vary with changes in production. (2.2.2)

Forwards Contracts to buy or sell on a fixed (future) date at a specified price. (5.2.2)

Future Value (FV) The value of an asset at some future time; see "Present Value." (3.2.1)

Futures Contracts to buy or sell at a future date at a price generally determined daily in an organized market. (5.2.2)

Income elasticity The percentage change in quantity demanded for one good or service in response to a one percent change in income. (2.1.3)

Independent System Operator (ISO) An operator of the transmission system that is not owned by any one user of the system. (5.2)

Inflation (i) An increase in the cost of purchasing a set of goods and services in comparison to the cost of the same set in a base period. (2.1.2)

Integrated Resource Planning (IRP) A methodology for determining a least-cost capacity expansion plan, considering all capacity resources, including Demand Side Management. (4.1.1)

Interest rate The rate of return required in the market on borrowed debt of a given risk class. (3.1)

Levelized capital cost An (annual) annuity so that the present value of the annuity is equal to the investment cost of the project. (3.2.1)

Liabilities An obligation to pay or perform a service. Liabilities appear on the balance sheet. (3.2.3)

Load profiling A technique for characterizing individual customers (loads) with typical usage patterns by their customer class. (5.4.2)

Long run A time such that no costs are fixed in the production process. Also, the "very long run" implies that technology is not fixed in the production process. (2.2.2)

Marginal Cost (MC) The change in total cost with a unit increase (or decrease) in production. (2.2.2)

Marginal Revenue (MR) The change in total revenue with a unit increase (or decrease) in sales. (2.2.2)

Market failure A situation in which a market is in equilibrium without price equal to marginal cost. These situations occur with (1) market (monopoly) power, (2) externalities, (3) public goods, and (4) asymmetric information. (2.2.3)

Market Operator (MO) An organization that arranges wholesale transactions that affect the flow of electricity in the transmission system. (5.2)

Market portfolio A combination of assets available in all markets. (Exercise 3.1)

Market power The ability to set output price above marginal cost. This ability can be legally sanctioned, as with intellectual property rights (e.g., patents), can be the result of a lack of competitors, or can be associated with positive economies of scale (leading to natural monopoly). (1.2)

Market-Clearing Price (MCP) The price that all sellers receive and that all buyers pay in a specific time-defined market. (5.2.1)

Mean An arithmetic average or the expected value of a variable. (3.4)

Monopoly A condition in which there is a single seller of a good or service. (2.1.1)

Natural monopoly A situation arising with a technology's positive economies of scale, such that a single firm can produce at the lowest cost.

Net Present Value (NPV) The discounted sum of costs (including investment) and benefits (including revenues). Calculating NPV for a project is generally equal to the present value of future cash flows minus the initial investment cost. (3.2.2)

Nodal pricing A transmission pricing method resulting in an energy price at every major transmission node (generator, transmission line junction, or substation) equal to the marginal cost of meeting an increase in demand at that node. (5.3.2)

Nominal prices Prices in the dollars of the year when a good or service is bought or sold. (2.1.2)

Opportunity cost The highest alternative value of all resources used in the production of a good or service. (2.2.1)

Peak-load pricing A pricing system whereby higher prices are charged during periods of high (electricity) consumption when the marginal cost of production is higher. (4.4.2)

Performance-Based Ratemaking (PBR) A regulatory pricing system whereby profit is tied to performance targets, rather than to costs; also known as incentive pricing. (4.3)

Perpetuity A type of annuity that pays a fixed payment forever. (3.2.1)

Poolco A structure in which all generators must sell all output to a single market. (5.2.1)

Present Value (*PV*) The value of future cash flows discounted to the present; also known as present discounted value. See "Net Present Value." (3.2.1)

Price discrimination The practice of charging different prices to different customers (with different price elasticities) or charging different prices for different amounts of the same good or service. (4.4.2)

Price elasticity The percentage change in quantity (demanded or supplied) associated with a one-percent change in price. (2.1.3)

Producer Surplus (*PS*) The difference between the market price and the variable cost of production, summed over all output. (2.3.1)

Profit (*PR*) The difference between total revenue and total cost. (2.3.1)

Qualified Generally referring to a retail customer who has the right to buy electricity from any retail electricity supplier. (1.5.4)

Rate of interest The rate of return required in the market on borrowed debt of a given risk class. (3.1)

Rate of return The percentage increase in the value of a firm's equity from one period to the next. (3.1)

Rate structure A set of tariffs charged for each type of service for each customer class. (4.2)

Rate-of-return regulation (ROR) A form of rate regulation in which the regulator compensates a firm for the cost of providing services to customers, including a rate of return on capital. (4.2)

Real prices Prices adjusted for the general level of inflation, measured relative to a constant set of goods and services through time. (2.1.2)

Real-time prices A pricing system whereby customers are charged the market price at the time of consumption for each unit they consume. (5.4.2)

Regulatory lag The period between two consecutive rate cases. It can also refer to the period between an unanticipated expenditure or change in revenue and the next rate case. (4.2)

Required Revenues (*RR*) The revenues required to cover the capital and operating costs of generating, transmitting, and distributing a regulated good or service. (4.1.1)

Retail choice The ability of end-users to choose their supplier; also known as customer choice. (5.4)

Retail price index (*RPI*) A measure of inflation used in the United Kingdom; similar to CPI. (2.1.2)

Revenue cap A system of regulation under which the utility can maximize profit subject to a maximum level of revenues. (4.3.2)

Risk The possibility of different outcomes given that the probability distribution of each outcome is known. (3.1)

Risk aversion The preference for a certain outcome over an uncertain outcome with the same expected value. (3.1)

Risk premium (*RP*) An amount such that the value of an uncertain outcome plus the risk premium is equal to the value of a certain outcome. (3.1)

Risk-free interest rate [nominal (R_f) and real (r_f)] A rate of interest that will be paid in all states of the world, proxied by the rate of interest on 90-day US government securities. (3.1)

Short run A period of production during which at least one input is fixed. (2.2.2)

Social surplus The sum of consumer surplus and producer surplus. (2.4)

Standard Deviation (*SD*) The square root of variance. (3.4)

Stocks Ownership shares in a firm's equity. (3.4.1)

Stranded assets The difference between the (book) value of a firm's assets under regulation and its (market) value under deregulation. (1.5.1)

Stranded costs The difference between a firm's required revenues under regulation and total cost under deregulation. (1.5.1)

Sunk cost A cost that cannot be recovered; for example, with the sale of an asset. (2.2.1)

Supply (*S*) The amount that producers are willing to sell at each price. (2.1.1)

Supply elasticity (E_s) The percentage change in quantity supplied associated with a one-percent change in price. (2.1.3)

System Operator (SO) The entity responsible for transmission system operation and reliability. (5.2)

Total Cost (*TC*) The total opportunity cost of producing a given level of output. (2.2.2)

Total Revenues (*TR*) The total revenue from selling a given level of output. (2.3.1)

Transmission congestion Congestion occurring when a transmission line or interface is not able to transmit more power because it is operating at its maximum transfer capacity. (5.3.3 and 6.4.2)

Two-part tariffs A pricing system in which the price is equal to (1) one part that allocates the fixed costs plus (2) another part that allocates the variable costs. (4.4.2)

Uncertainty A situation in which the probability of an outcome is unknown; compare with "Risk." (3.1)

Uniform pricing A transmission pricing method in which there are no energy price differences within the transmission network. (5.3.2)

Utility function (*U*) A function that orders a consumer's preferences from most preferred to least preferred. Under specific assumptions, the utility function can assign cardinal values to represent levels of preference. (Exercise 3.2)

Variable Cost (*VC*) Costs that vary with changes in the level of production. (2.2.2)

Variance (*VAR*) A measure of dispersion of a sample, equal to a weighted sum of the differences of each observation from the mean of the population. (3.4)

Weighted Average Cost of Capital (*WACC*) A weighted average of the firm's cost of debt and cost of equity. (3.4.2)

Zonal pricing A transmission pricing method in which energy prices vary by zone, where zones are defined by sets of nodes where it is unlikely that the network will experience congestion. Congestion can appear at interfaces between zones. (5.3.2)

REFERENCES

Arraiza J. M. 1998. Los costes de transición a la competencia (Competitive transition costs). *Revista Anales de Mecánica y Electricidad.* Número Monográfico sobre la Liberalización del Sector Eléctrico Español (May).

Averch, H. and L. Johnson. 1962. Behavior of the firm under regulatory constraint. *American Economic Review 52*(5): 1052–1069.

Barker, J., B. Tenenbaum, and F. Wolf. 1997. Governance and regulation of power pools and system operators: An international comparison. Technical working paper, World Bank, no. 382.

Bastos, C. and M. Abdala. 1996. *Reform of the electric power sector in Argentina.* Buenos Aires, Argentina: Editorial Antártica.

Berg, S. and J. Tschirhart. 1988. *Natural monopoly regulation.* Cambridge: Cambridge University Press.

Besant-Jones, J. and B. Tenenbaum. 2001. *The California power crisis: Lessons for developing countries.* The World Bank, ESMAP Report (May).

Blumstein, C. and J. Bushnell. 1994. A guide to the Blue Book: Issues in California's electric industry restructuring and reform. *The Electricity Journal 7*(7): 18–29.

BOE (Boletín Oficial del Estado). 1997a. *Ley 54/1997, de 27 de Noviembre, del Sector Eléctrico* (Electricity Law 54/1997). Madrid, Spain: BOE no. 285.

BOE (Boletín Oficial del Estado). 1997b. *Real Decreto 2017/1997, de 26 de Diciembre, por el que se organiza y regula el procedimiento de liquidación de los costes del transporte, distribución y comercialización a tarifa, de los costes permanentes del sistema y de los costes de diversificación y seguridad de abastecimiento* (Royal decree on the general settlement procedure for the regulated activities). Madrid, Spain.

BOE (Boletín Oficial del Estado). 1997c. *Real Decreto 2019/1997, de 26 de Diciembre, por el que se organiza y regula el mercado de producción de energía eléctrica* (Royal decree on the organization and regulation of the wholesale electricity market). Madrid, Spain.

BOE (Boletín Oficial del Estado). 1998. *Real Decreto 2028/1998, de 23 de Diciembre, por el que se establecen tarifas de acceso a las redes* (Royal decree on access tariffs to the networks). Madrid, Spain.

BOE (Boletín Oficial del Estado). 1999. *Real Decreto 2066/1999, de 30 de Diciembre, por el que se establece la tarifa eléctrica para el 2000* (Royal decree setting electricity tariffs for the year 2000). Madrid, Spain.

BOE (Boletín Oficial del Estado). 2000a. *Real Decreto—ley 6/2000, de 23 de Junio, de Medidas Urgentes de Intensificación de la Competencia en Mercados de Bienes y Servicios*

(Royal decree on urgent actions to intensify competition in commodity and service markets). Madrid, Spain.

BOE (Boletín Oficial del Estado). 2000b. *Real Decreto 1955/2000, de 1 de Diciembre, por el que se regulan las actividades de transporte, distribución, comercialización y suministro y los procedimientos de autorización de instalaciones de energía eléctrica* (Royal decree on transmission, distribution, retail, and supply of the electrical energy). Madrid, Spain.

Boiteux, M. 1960. Peak load pricing. *Journal of Business 33:* 157–179.

Borenstein, S. 2001. The trouble with electricity markets (and some solutions). Working paper, UC Energy Institute, Berkeley, California (January): PWP-081.

Borenstein, S., J. Bushnell, and F. Wolak. 2000. Diagnosing market power in California's deregulated wholesale electricity market. Working paper, UC Energy Institute, Berkeley, California (July): PWP-064. www.stanford.edu/~wolak.

Borenstein, S., J. Bushnell, E. Kahn, and S. Stoft. 1995. Market power in California electricity markets. *Utilities Policy 5*(3/4): 219–236.

Brealey, R., and S. Myers. 2000. *Principles of corporate finance.* 6th ed. McGraw-Hill.

Bushnell, J. and S. Oren. 1997. Transmission pricing in California's proposed electricity market. *Utilities Policy 6*(3): 237–244.

California Power Exchange. 1998. *PX primer: California's new electricity market.* Ver. 3 (March).

California Power Exchange. 1999. *Block forwards market background.* (June).

CAMMESA (Compañía Administradora del Mercado Eléctrico Mayorista SA) 2000. *Procedimientos para la programación de la operación, el despacho de cargas y el calculo de precios.* Vol. 15 (May). Buenos Aires, Argentina.

CENELEC (European Committee for Electrotechnical Standardization). 1994. *Voltage Characteristics of Electricity Supplied by Public Distribution Systems.* European Norm EN 50160 (November).

Che, Y.K. and G.S. Rothwell. 1995. Performance-based pricing for nuclear power plants. *The Energy Journal 16*(4): 57–77

Christensen, L. and W. Greene. 1976. Economies of scale in U.S. electric power generation. *Journal of Political Economy 84*(4): 655–676.

Christie, R., B. Wollenberg, and I. Wangensteen. 2000. Transmission management in the deregulated environment. *Proceedings of the IEEE 88*(2): 170–195.

CNE (Comisión Nacional de Energía). 2000. *El Consumo Eléctrico en el Mercado Peninsular en 1999* (Electricity demand in 1999) (July). Madrid, Spain.

CNSE (Comisión Nacional del Sistema Eléctrico). 1995. *Atlas de la Distribución* (Distribution data). Madrid, Spain.

CNSE (Comisión Nacional del Sistema Eléctrico). 1998. *The new Spanish Electricity Act and the introduction of a competitive electricity market in Spain* (in Spanish). Madrid, Spain (February).

Cobb, C. and P. Douglas. 1928. A theory of production. *American Economic Review 18*(Supp.): 139–165.

Comnes, A., S. Stoft, N. Greene, and L. Hill. 1995. *Performance-based ratemaking for electric utilities: A review of plans and analysis of economic and resource-planning issues,* Vol. 1. Lawrence Berkeley National Laboratory, Berkeley, California (November): LBNL-37577.

CPUC (California Public Utilities Commission). 1993. *California's electric services industry: Perspectives on the past, strategies for the future* ("Yellow Paper"). San Francisco, California: Division of Strategic Planning (February).

CPUC (California Public Utilities Commission). 1994. *Order instituting rulemaking and order instituting investigation* (the "Blue Book"). San Francisco, California (April): R.94-04-031 and I.94-04-032.

CPUC (California Public Utilities Commission). 1995. *Proposed policy decision adopting a preferred industry structure.* San Francisco, California (May): R.94-04-031 and I.94-04-032.

CPUC (California Public Utilities Commission). 1997a. *Decision 97-05-040.* San Francisco, California.

CPUC (California Public Utilities Commission). 1997b. *Decision 97-05-039.* San Francisco, California.

CPUC (California Public Utilities Commission). 1997c. *Electric and gas utility performance based Ratemaking mechanisms.* Prepared by R. Myers and A. Johnson. San Francisco, California (December).

CPUC (California Public Utilities Commission). 1997d. *Decision 97-12-048.* San Francisco, California.

Crew, M., C. Fernando, and P. Kleindorfer. 1998. The theory of peak-load pricing: A survey. In *The foundations of regulatory economics.* Edited by R. Ekelund. London: Elgar.

Detroit Edison. 1998. Draft: Detroit Edison customer choice plan, submitted to the staff of the Michigan Public Service Commission (April 6).

DISGRUP (Grupo de Trabajo de Distribución y Comercialización). 1998. *Propuesta de reglamento de la regulación de distribución y comercialización de energía eléctrica* (Proposal for the regulation of distribution and retail supply of electricity). Madrid, Spain (March).

Distribution Loss Factors Working Group. 1998. *Distribution loss factors working group report.* Submitted to the California Public Utilities Commission, San Francisco, California: R.94-04-031 and I.94-04-032.

Dixit, A. 1992. Investment and hysteresis. *Journal of Economic Perspectives 6*(1):107–132.

Dixit, A. and R. Pindyck. 1994. *Investment under uncertainty.* Princeton: Princeton University Press.

DOJ/FTC (Department of Justice/Federal Trade Commission). 1992. *Horizontal merger guidelines.* Washington, DC.

ECON. 1999. Measuring the efficiency of Statnett. Oslo, Norway: Centre for Economic Analysis AS: Report 25/99.

EDF (Electricité de France)—DISTRELEC. 1989. *Tarifas. Informe de diagnóstico y orientación. Proyecto de ingeniería. Sistemas de distribución.* Buenos Aires, Argentina.

EIA (Energy Information Administration). 1993. *The Public Utility Holding Company Act of 1935: 1935–1992.* DOE/EIA-0563.

EIA (Energy Information Administration). 1995. *Annual electric utility report 1994.* Form EIA-861.

EIA (Energy Information Administration). 1996. *The changing structure of the electric power industry: An update.* DOE/EIA-0562 (96).

EIA (Energy Information Administration). 1997. *Financial statistics of major U.S. investor-owned electric utilities 1996.* DOE/EIA-0437(96)/1.

EIA (Energy Information Administration). 1998. *The changing structure of the electric power industry: Selected issues.* DOE/EIA-0620.

EIA (Energy Information Administration). 1999a. *Electric power annual 1998.* Vol. 1. DOE/EIA-0348(98)/1.

EIA (Energy Information Administration). 1999b. *Electric sales and revenue 1998.* DOE/EIA-0540(98)/1.

EIA (Energy Information Administration). 1999c. *Annual electric utility report 1998.* Form EIA-861.

EIA (Energy Information Administration). 1999d. *Monthly electric utility sales and revenue report with state distributions.* Form EIA-826.

Electricity Law. 1992. *Ley No. 24.065 de Energia Electrica.* Buenos Aires, Argentina. (January).

ENEL (Ente Nacional de l'Energia)—DISTRELEC. 1989. *Estudio del costo de falla en el suministro de energía eléctrica. Proyecto de ingeniería. Sistemas de distribución.* Buenos Aires, Argentina.

Energy Act of 1990: Act relating to generation, conversion, transmission, trading, and distribution of energy, etc. (in Norwegian). 1990. Oslo, Norway: Norwegian Ministry of Petroleum and Energy. www.lovdata.no

Energy Policy Act of 1992 (EPAct). 1992. U.S. Public Law 102-486.

European Parliament and Council. 1996. *The internal electricity market directive.* Brussels, Belgium (December): Directive 96/92/EC.

FERC (Federal Energy Regulatory Commission). 1996a. *Order No. 888* (Docket Nos. RM95-8-000, Promoting wholesale competition through open access nondiscriminatory transmission services by public utilities, and RM94-7-001, recovery of stranded costs by public utilities and transmitting utilities) (April).

FERC (Federal Energy Regulatory Commission). 1996b. *Order No. 889* (Docket Nos. RM95-9-000, Open access same-time information system, formerly Real-Time information networks and standards of conduct) (April).

FTC (Federal Trade Commission). 1996. In the matter of inquiry concerning (the Federal Energy Regulatory) Commission's merger policy under the Federal Power Act. www.ftc.gov/be

Fox-Penner, P. 1997. *Electric utility restructuring: A guide to the competitive era.* Public Utilities Reports.

Gedra, T. (forthcoming). *Power economics and regulation.* Norwell, MA: Kluwer Academic Publishers.

Gilbert, R. and E. Kahn (eds.). 1996. *International comparisons of electricity regulation.* Cambridge: Cambridge University Press.

Golove, W., R. Prudencio, R. Wiser, and C. Goldman. 2000. Electricity restructuring and value-added services: Beyond the hype. Presented at the American Council for an Energy-Efficient Economy 2000 Summer Study, 20-25 August, Asilomar, California.

Gómez, T., C. Marnay, A. Siddiqui, L. Liew, and M. Khavkin. 1999. *Ancillary services markets in California.* Lawrence Berkeley National Laboratory, Berkeley, California: LBNL-43986.

Grande, O. and I. Wangensteen. 2000. Alternative models for congestion management and pricing: Impact on network planning and physical operation. Presented at CIGRE (International Council on Large Electric Systems) 2000, 27 August–1 September, Paris, France. www.cigre.org

Grasto, K. 1997. Incentive-based regulation of electricity monopolies in Norway. Norwegian Water Resources and Energy. www.nve.no

Green, R. 1999. Draining the pool: The reform of electricity trading in England and Wales. *Energy Policy 27*(9): 515–525.

Green, R. 2000. Competition in generation: The economic foundations. *Proceedings of the IEEE 88*(2): 128–139.

Grønli, H., T. Gómez, and C. Marnay. 1999. Transmission grid access and pricing in Norway, Spain, and California—A comparative study. Presented at Power Delivery Europe 1999, 28–30 September, Madrid, Spain.

Guidi, C. 1997. Estudio de los costes de la distribución y cargos por uso de la red en Argentina. Internal report, Instituto de Investigación Tecnológica, Madrid, Spain. (December).

Hall, D. 1998. *Electric utility cost exercises.* California State University, Long Beach (September).

Hall, D. and J. Hall. 1994. *Evaluation of alternative electricity systems.* California State University, Long Beach (June).

Hausker, K. 1993. Two cheers for the Energy Policy Act! *The Electricity Journal 6*(1): 26–32.

Hogan, W. 1992. Contract networks for electric power transmission. *Journal of Regulatory Economics 4*(3): 211–242.

Hunt, S. and G. Shuttleworth. 1996. *Competition and choice in electricity.* New York: Wiley.

Ilic, M., F. Galiana, and L. Fink (eds.). 1998. *Power system restructuring: Engineering and economics.* Norwell, MA: Kluwer Academic Publishers.

ISO (Independent System Operator). 1998. *California Independent System Operator operating agreement and tariff.* Folsom, California.

ISO (Independent System Operator). 1999a. *Report on impacts of RMR contracts on market performance.* Folsom, California (March). www.caiso.com

ISO (Independent System Operator). 1999b. *Annual report on market issues and performance.* Folsom, California (June). www.caiso.com

ISO/PX (Independent System Operator/Power Exchange). 1997. *Phase II Filings of the California Power Exchange Corporation and Phase II Filings of the California Independent System Operator Corporation to the Federal Energy Regulatory Commission* (March): Docket No. EC96-19-001 and EC96-1663-001.

Jordanger, E. and H. Grønli (eds.) 2000. Deregulation of the electricity supply industry: Norwegian experience 1991–2000. Trondheim, Norway: SINTEF Energy Research (October): TR A5285.

Joskow, P. 1997. Restructuring, competition and regulatory reform in the US electricity sector. *Journal of Economic Perspectives 11*(3): 119–138.

Joskow, P. 2001. California's electricity crisis. NBER Working Paper 8442. www.nber.org/papers/w8442

Joskow, P. and E. Kahn. 2000. *A quantitative analysis of pricing behavior in California's wholesale electricity market during summer 2000.* Prepared for Southern California Edison (November).

Joskow, P. and R. Noll. 1981. Regulation in theory and practice: An overview. In *Studies in public regulation.* Edited by G. Fromm. Cambridge: MIT Press.

Joskow, P. and R. Schmalensee. 1983. *Markets for power: An analysis of electric utility deregulation.* Cambridge: MIT Press.

Joskow, P. and R. Schmalensee. 1986. Incentive regulation for electric utilities. *Yale Journal on Regulation 4*(1): 1–49.

Kahn, A. 1988. *The economics of regulation, principals, and institutions.* Cambridge: MIT Press.

Kahn, E. 1991. *Electric utility planning and regulation.* 2nd ed. Washington, DC: American Council for an Energy-Efficient Economy.

Kahn, E. 1995. The electricity industry in Spain. Working paper, UC Energy Institute, Berkeley, California (July): PWP-032.

Kaminski, V. 1997. The challenge of pricing and risk managing electricity derivatives. Chap. 10 in *The U.S. power market.* London: Risk Publications.

Kittelsen, S. 1993. Stepwise DEA: Choosing variables for measuring technical efficiency in Norwegian electricity distribution. Bergen, Norway: Foundation for Research in Economics and Business Administration (SNF): A55/93.

Laffont, J.J. and J. Tirole. 1993. *A theory of incentives in procurement and regulation.* Cambridge: MIT Press.

Langset, T. and A. Torgersen. 1997. Effektivitet i distribusjonsnettene 1995. Oslo, Norway: Norwegian Water Resources and Energy: Report 1997/15.

Littlechild, S. and M. Beesley. 1989. The regulation of privatized monopolies in the United Kingdom. *Rand Journal of Economics 20*(3): 454–72.

Livik, K. and S. Fretheim (eds.) 1997. *Deregulation of the Nordic power market: Implementation and experiences, 1991–1997.* Trondheim, Norway: SINTEF Energy Research (November): TR A4602.

Marnay, C. and S. Pickle. 1998. Power supply expansion and the nuclear option in Poland. *Contemporary Economic Policy 16*(1): 109–121.

Marcus, W. and J. Hamrin. 2001. How we got into the California energy crisis. White paper, Center for Resource Solutions and JBS Energy, Inc. Sacramento, California.

Marmolejo, A. and S. Williams. 1995. *The Argentine power book. Latin America research—Electric utilities.* Frankfurt: Keinwort Benson Research.

McCullough, R. 2001. Price spike tsunami: How market power soaked California. *Public Utilities Fortnightly 139*(1): 22–32.

MEOSP (Ministerio de Economía y Obras y Servicios Públicos). 1992. *Contrato de concesión de los servicios de distribución y comercialización.* Buenos Aires, Argentina.

Moore, I. and J. Anderson. 1997. *Introduction to the California power market* (July).

Navarro, P. 1996. Seven basic rules for the PBR regulator. *The Electricity Journal 9*(3): 24–30.

NERA (National Economic Research Associates). 1998. *Analysis of the reform of the Argentine power sector: Final report.* Buenos Aires, Argentina.

Nerlove, M. 1963. Returns to scale in electricity supply. In *Measurement in econometrics.* Edited by C. Christ. Stanford, CA: Stanford University Press.

Nord Pool (Nordic Power Exchange). 1998. *The Elspot market. The spot market.* Nord Pool (October). www.nordpool.no

NORDEL (Nordic Organization for Electric Cooperation). 2001. *NORDEL Statistics 1999.* www.nordel.org

NVE (Norwegian Water Resources and Energy Directorate). 1999. Regulations concerning financial and technical reporting, permitted income for network operations and transmission tariffs. Oslo, Norway: Norwegian Water Resources and Energy. www.nve.no

OED (Ministry of Petroleum and Energy). 2000. Faktahefte: Energi—og vassdragsvirksomheten i Norge. Oslo, Norway: www.dep.no/oed/norsk/publ/veiledninger/026021-120002/index-dok000-b-n-a.html

OMEL (Compañía Operadora del Mercado Español de Electricidad). 1998a. *Reglas de funcionamiento del mercado de producción de energía eléctrica* (Rules for running the wholesale generation electricity market). Madrid, Spain.

OMEL (Compañía Operadora del Mercado Español de Electricidad). 1998b. *Mercado de Electricidad. Evolución del mercado de producción de energía eléctrica* (Evolution of the electricity wholesale market). (December).

OMEL (Compañía Operadora del Mercado Español de Electricidad). 1999. *Mercado de Electricidad. Evolución del mercado de producción de energía eléctrica* (Evolution of the electricity wholesale market). (December).

Oren, S. 2000. Capacity payments and supply adequacy in competitive electricity markets. Presented at the VII Symposium of Specialists in Electric Operational and Expansion Planning (SEPOPE), 21–26 May, Curitiba, Brazil.

Pasqualetti, M. and G. Rothwell (eds.) 1991. *Nuclear decommissioning economics,* a special issue of *The Energy Journal 12.*

Patrick, R. and F. Wolak. 1999. Customer response to real-time pricing in the England and Wales electricity market: Implications for demand-side bidding and pricing options design under competition. In *Regulation under increasing competition.* Edited by M. Crew. Norwell, MA: Kluwer Academic.

Pérez-Arriaga, J. I. 1994. *The organization and operation of the electricity supply industry in Argentina.* London: Energy Economic Engineering Ltd.

Pérez-Arriaga, J. I. 1997. *The competitive electricity market under the new Spanish Act.* Presented by the Spanish Comisión Nacional del Sistema Eléctrico (CNSE) in Warsaw, Poland (December).

Pérez-Arriaga, J. I. 1998. Visión global del cambio de regulación (Regulatory changes: A general overview). *Revista Anales de Mecánica y Electricidad.* Número Monográfico sobre la Liberalización del Sector Eléctrico Español.

Pickle, S., C. Marnay, and F. Olken. 1997. Information systems requirements for a deregulated electric power industry. *Utilities Policy 6*(2): 163–176.

Pindyck, R. and D. Rubinfeld. 2001. *Microeconomics.* 5th ed. Upper Saddle River, NJ: Prentice Hall.

PURPA (Public Utility Regulatory Policies Act of 1978). U.S. Public Law 95-617.

Ramsey, F. 1927. A contribution to the theory of taxation. *Economic Journal 37*(145): 47–61.

Román J., T. Gómez, A. Muñoz, J. Peco. 1999. Regulation of distribution network business. *IEEE Transactions on Power Delivery 14*(2): 662–669.

Rothwell, G. 1997. Continued operation or closure: the net present value of nuclear power plants. *The Electricity Journal* (Aug./Sept.): 41–48.

Rothwell, G. 2000. The risk of early retirement of US nuclear power plants under electricity deregulation and carbon dioxide reductions. *The Energy Journal 21*(3): 61–87

Rothwell, G. 2001. Probability distributions of net present values for U.S. nuclear power plants. *The Utilities Project.* Montgomery Research, San Francisco, California. www.UtilitiesProject.com

Rothwell, G. and K. Eastman. 1987. A note on allowed and realized rates of return in the electric utility industry. *Journal of Industrial Economics 36*(1): 105–110.

Rothwell, G., J. Sowinski, and D. Shirey. 1995. Electric utility demand-side management in Poland. *Contemporary Economic Policy* (Jan.): 84–91.

Rothwell, G. and J. Sowinski. 1999. A real options approach to investment planning in electricity supply. Proceedings of the Electricity Markets Conference, 14–15 May, Naleczow, Poland.

Schwarz, J., K. Staschus, T. Knop, and K. Zettler. 2000. Overview of the EU electricity directive. *IEEE Power Engineering Review 20*(4): 4–7.

Schweppe, F., M. Caramanis, R. Tabors, and R. Bohn. 1988. *Spot pricing of electricity.* Norwell, MA: Kluwer Academic Publishers.

Shirmohammadi, D. and P. Gribik. 1998. *Zonal market clearing prices: A tutorial.* Prepared for the California Power Exchange (March).

Siddiqui, A., C. Marnay, and M. Khavkin. 2000. Excessive price volatility in the California ancillary services markets: Causes, effects, and solutions. *The Electricity Journal 13*(6): 58–68.

Spiller P.T. and L.V. Martorell. 1996. How should it be done? Electricity regulation in Argentina, Brazil, Uruguay, and Chile. In *International comparisons of electricity regulation.* Edited by R. Gilbert and E. Kahn. Cambridge: Cambridge University Press.

SSB (Statistics Norway). 2001. *Statistical yearbook of Norway 2000.* Statistics Norway. www.ssb.no/english/yearbook

Stoft, S. 1996. California's ISO: Why not clear the market? *The Electricity Journal 9*(10): 38–43.

Stoft, S. 1998. Congestion pricing with fewer prices than zones. *The Electricity Journal. 11*(4): 23–31.

Stoft, S. 2002. *Power system economics: Designing markets for electricity.* New York: IEEE Press. www.stoft.com

Stoft, S., T. Belden, C. Goldman, and S. Pickle. 1998. *Primer on electricity futures and other derivatives.* Lawrence Berkeley National Laboratory, Berkeley, California (January): LBNL-41098.

Stoll, H.G. 1989. *Least-cost electric utility planning.* New York: Wiley.

Sweeney, J. 2002. *The California electricity crisis.* Stanford, CA: Hoover Press.

Takayama, A. 1993. *Analytical methods in economics.* Ann Arbor: University of Michigan Press.

Unda, J.I. 1998. Liberalization of the Spanish electricity sector: An advanced model. *The Electricity Journal 11*(5): 29–37.

UNESA (Asociación Española de la Industria Eléctrica). 1998. *La Tarifa Eléctrica 1998.* (The Electricity Tariff of 1998). Madrid, Spain (March).

Varian, H. 1992. *Microeconomic analysis.* 3rd ed. New York: Norton.

Vazquez, C., M. Rivier, and I.J. Pérez-Arriaga. 2002. A market approach to long-term security of supply. *IEEE Transactions on Power Systems 17*(2): 349–357.

Viscusi, W.K., J. Vernon, and J. Harrington. 2000. *Economics of regulation and antitrust.* 3rd ed. Cambridge: MIT Press.

Voldhaug, L., T. Granli, and S. Bygdås. 1998. Reliability dependent service pricing by means of outage compensation to customers. Presented at Power Delivery Europe 1998, 23–26 October, London.

Watkiss, J. and Smith, D. 1993. The Energy Policy Act of 1992: A watershed for competition in the wholesale power market. *Yale Journal on Regulation 10*(2): 447–492.

WEPEX (Western Power Exchange). 1996. Joint application of the Pacific Gas and Electric Company, San Diego Gas and Electric Company, and Southern California Edison Company to sell electric energy at market-based rates using a power exchange. Filing before the U.S. Federal Energy Regulatory Commission.

Wiser, R., W. Golove, and S. Pickle. 1998. California's electric market: What's in it for the customer. *Public Utilities Fortnightly 136*(15): 38–45.

Wolak, F., R. Nordhaus, and C. Shapiro. 1998. *Preliminary report on the operation of the ancillary services markets of the California Independent System Operator.* Market Surveillance Committee of the California ISO (August). www.caiso.com

Woolf, T., and J. Michals. 1995. Performance-based ratemaking: Opportunities and risks in a competitive electricity industry. *The Electricity Journal 8*(8): 64–73.

WEB SITES

Argentina Ministerio de Economía: www.mecon.gov.ar

Australia—Independent Pricing and Regulatory Tribunal (IPART) of New South Wales: www.ipart.nsw.gov.au

Australia—TRANSGRID: www.tg.nsw.gov.au

California Energy Commission: www.energy.ca.gov

California Independent System Operator: www.caiso.com

California Public Utilities Commission: www.cpuc.ca.gov

CAMMESA (Compañía Administradora del Mercado Mayorista Eléctrico SA). Argentina: www.cammesa.com.ar

Comisión Nacional de Energía (Spanish Regulator): www.cne.es

Compañía Operadora del Mercado Español de Electricidad (Spanish Market Operator): www.omel.es

Denmark—Elkraft System: www.elkraft.dk

Denmark—Eltra: www.eltra.dk

EIA (US Department of Energy, Energy Information Administration): www.eia.doe.gov

ENRE (Ente Nacional Regulador de la Electricidad) de Argentina: www.enre.gov.ar

ESSA (Electricity Supply Association of Australia): www.esaa.com.au

European Commission: www.europa.eu.int/comm

FERC (U.S. Federal Energy Regulatory Commission): www.ferc.fed.us

Finland—Fingrid (Finish grid operator): www.fingrid.fi

Maryland Public Service Commission, US: www.psc.state.md.us

Ministerio de Ciencia y Tecnología—Energía y Minas (Ministry of Science and Technology-Energy and Mines): www.mcyt.es

NARUC (National Association of Regulatory Utility Commissioners) US: www.naruc.org

New England ISO, U.S.: www.iso-ne.com
New York Independent System Operator, US: www.nyiso.com
NORDEL (Nordic Organization for Electric Cooperation): www.nordel.org
NORDPOOL (Nordic Power Exchange): www.nordpool.no
North American Electric Reliability Council: www.nerc.com
NVE (Norwegian Water Resources and Energy Directorate): www.nve.no
OFGEM (Office of Gas and Electricity Markets), UK: www.ofgem.gov.uk
Office of the Regulator—General, Victoria, Australia: www.reggen.vic.gov.au
Pennsylvania Public Utility Commission: www.puc.paonline.com
PJM (Pennsylvania New Jersey Maryland) interconnection: www.pjm.com
Red Eléctrica de España (Spanish System Operator & Transmission Owner): www.ree.es
Statnett, Norway: www.statnett.no
UNESA (Asociación Española de la Industria): www.unesa.es
UK Electricity Association: www.electricity.org.uk

AUTHOR INDEX

SUBJECT INDEX

ABOUT THE AUTHORS

Geoffrey Rothwell is the Director of Honors Programs for the Department of Economics and is Associate Director of the Public Policy Program at Stanford University, Stanford, California. He received his Masters degree in Jurisprudence and Social Policy from Boalt Law School in 1984 and his Ph.D. in Economics from the University of California, Berkeley, in 1985. He was a Post-Doctoral Fellow at the California Institute of Technology from 1985 to 1986. Teaching at Stanford since 1986, Dr. Rothwell is widely published in the economics of electricity and nuclear power, including nuclear fuel markets, nuclear power plant construction, operating costs, productivity, reliability, decommissioning, and spent nuclear fuel management. From 1995 to 1997, he chaired the Committee on Methodology for Nuclear Power Plant Performance and Statistical Analysis of the International Atomic Energy Agency, Vienna. Since 2001, he has served on the U.S. Department of Energy's Generation IV Nuclear Energy Systems Roadmap Committee and is currently serving on the Economic Models Working Group. Publications include analyses of nuclear power industries in China, France, Russia, the United Kingdom, and the United States. Dr. Rothwell began working in Russia in 1992 with the Russian Academy of Sciences, and is now affiliated with the New Economics School in Moscow, working on a project to evaluate market reforms in the Russian electric utility sector.

Tomás Gómez San Román is a professor of Electrical Engineering at the Engineering School of Universidad Pontifica Comillas (UPCo) in Madrid, Spain. He obtained the Degree of Doctor Ingeniero Industrial from Universidad Politécnica, Madrid in 1989, and the Degree of Ingeniero Industrial in Electrical Engineering from UPCo in 1982. He joined Instituto de Investigación Tecnológica at UPCo (IIT-UPCo) in 1984. From 1994 to 2000, he was the Director of IIT, and from 2000 to 2002, the Vice-Rector of Research, Development, and Innovation of UPCo. Dr. Gómez has vast experience in industry joint research projects in the field of Electric Energy Systems with Spanish, Latin American, and European utilities. He has been project manager and/or principal investigator for more than 40 research projects.

Areas of interest include operation and planning of transmission and distribution systems, power quality assessment and regulation, and economic and regulatory issues in the electrical power sector. He has published more than 50 articles in different specialized magazines such as IEEE PES Transactions and Conference proceedings. He is a member of IEEE and belongs or has belonged to the Technical Committees of the Conferences: Probabilistic Methods Applied to Power Systems, Power System Computation Conference, and IEEE Power Technology. From 1998 to 1999, he was a visiting researcher at the Energy Analysis Department of the Lawrence Berkeley National Laboratory, California.